RF Power Semiconductor Generator Application in Heating and Energy Utilization

Satoshi Horikoshi · Nick Serpone
Editors

RF Power Semiconductor Generator Application in Heating and Energy Utilization

Editors
Satoshi Horikoshi
Department of Materials and Life Science
Sophia University
Chiyodaku, Tokyo, Japan

Nick Serpone
Dipartimento di Chimica
Universita di Pavia
Pavia, Italy

ISBN 978-981-15-3547-5 ISBN 978-981-15-3548-2 (eBook)
https://doi.org/10.1007/978-981-15-3548-2

This Springer imprint is published by the registered company Springer Nature Singapore Pte Ltd.
The registered company address is: 152 Beach Road, #21-01/04 Gateway East, Singapore 189721, Singapore

Preface

In 1946, J. P. Eckert and J. Mauchly invented the Electronic Numerical Integrator and Computer (ENIAC) registered as the world's earliest electronic calculator, which used 17,468 vacuum tubes powered by 150 kW and running calculations at 5000 times per second. Although the size and power consumption are unthinkable at present, the computing speed was nothing less than amazing for a computer in the mid-1940s. By comparison, a current personal computer (with an Intel Core i7) is capable of performing some 5.3×10^9 calculations per second (i.e., 5,300,000,000 or 5.3 trillion calculations per second) with a power consumption of only about 20–30 W. Clearly, a current PC has evolved into a different dimension compared to the ENIAC. So what caused this evolution?

We are reminded that the innovative evolution of primitive organisms was due to the introduction of mitochondria into the body and the success of transcending time and space limitations. In the case of computers, the brain evolved from a vacuum tube technology to a semiconductor technology. This *conversion from vacuum tube devices to semiconductor devices* is not limited solely to the computer field; in recent years, it is also seen in other fields. For instance, switching from a cathode ray tube (vacuum tube) TV to a liquid crystal (semiconductor control) TV has afforded TVs with an incredible increase in screen size. Analogous to the above, the switch from fluorescent lamps (vacuum tube lamps) to LEDs (semiconductor lamps) has also witnessed significant progress owing to concerns regarding mercury-related health issues. Although we no longer see common products that use vacuum tubes, it is worth noting that magnetrons (that is, microwave oscillators operating in microwave ovens) still use vacuum tube technology. Such ovens may well be the last vacuum tube products.

Recent years have witnessed the appearance of an increasing number of technical books that explain power semiconductors, while but scant specialized books are available in the heating and energy fields that make use of RF semiconductor generators. In response to this need, the current book examines and explains the use of RF power semiconductor generators in heating applications, energy applications, and even new applications. In particular, features are presented that explain the beneficial use of RF power semiconductor oscillators in the heating and energy fields vis-à-vis conventional heating and energy application technologies. The book also covers a wide range of fields that include flow chemistry, curing of adhesives and resins, cooking ovens, RF discharge lamps, plasma, automobiles, wireless transmission, and magnetic recording. As such, the co-editors hope this book will provide implicit hints to microwave researchers for yet to be discovered notions and novel ideas to researchers in various other fields.

We are grateful to all the contributing authors who have answered our call, and thank the Springer editorial staff for their thorough and professional assistance. The data presented would not have been possible without the fruitful collaboration of many university and industrial researchers, and not least without the cooperation of students whose names appear in many of the publications cited. We are indeed very grateful for their efforts.

From Midsummer Tokyo
August 2019

Satoshi Horikoshi
Nick Serpone

Contents

About the Editors

Satoshi Horikoshi Professor, Sophia University, Department of Materials and Life Sciences, email: horikosi@sophia.ac.jp.

Satoshi Horikoshi received his PhD degree in 1999 and was subsequently a postdoctoral researcher at the Frontier Research Center for the Global Environment Science (Ministry of Education, Culture, Sports, Science and Technology) until 2006. He joined Sophia University as Assistant Professor in 2006 and then moved to Tokyo University of Science as Associate Professor in 2008, after which he returned to Sophia University as Associate Professor in 2011 and Professor in 2020. Currently, he is on the Editorial Advisory Board of the Journal of Microwave Power and Electromagnetic Energy and other international journals . His research interests involve new functional material, nanomaterial synthesis, molecular biology, formation of sustainable energy, environmental protection using microwave- and/or photo-energy. He has co-authored over 190 scientific publications and has contributed to and edited or co-edited 27 books. He frequently explains microwaves on television and in newspapers.

Nick Serpone, Ph.D., F. EurASc. Visiting Professor, PhotoGreen Laboratory, Dipartimento di Chimica, Universita di Pavia, Italia, email: nick.serpone@unipv.it; Postal address: 5647 Smart Avenue, Cote Saint-Luc (QC), H4W-2M4, Canada.

Nick Serpone is Professor Emeritus (Concordia University, Montreal, Canada) and since 2002 has been a Visiting Professor at the University of Pavia (Italy). He was also a Visiting Professor at the Universities of Bologna and Ferrara (Italy), École Polytechnique Fédérale de Lausanne (Switzerland), École Centrale de Lyon (France), Tokyo University of Science (Japan), and Guest Lecturer at the University of Milan (Italy). He was the IBO Chemistry Program Director at the National Science Foundation (Arlington, VA, USA; 1998–2001) and consultant to the 3M Company (USA; 1986–1996). He has co-edited/co-authored several books (12), contributed 34 chapters to books, and has published over 480 articles. His principal interests have focused on the photophysics and photochemistry of coordination compounds and metal-oxide semiconductors, photocatalysis for environmental remediation, and microwave-assisted chemistry. He was elected a Fellow of the European Academy of Sciences in 2010 (EurASc) and is currently Head of the Materials Science Division of EurASc (2014–2020).

Part I
Solid State RF

Chapter 1
RF Energy System with Solid State Device

Naoki Shinohara

Abstract Radio frequency (RF) energy is generated from electricity via either a vacuum tube or a solid state device. Owing to recent advances in solid state device technology, high power amplifiers can be applied in microwave ovens. A wide band gap material like gallium nitride (GaN) is expected to be a good candidate for high power RF energy applications. GaN is usually applied as a blue light-emitting diode but can also be employed for power devices (e.g., converter). Historically, in contrast to wide band gap semiconductor-based devices, conventional silicon (Si)-based solid state devices were considered incompatible for high power applications. Recently, however, Si-based laterally diffused metal–oxide semiconductors (LDMOSs) have been applied successfully in RF energy systems. When solid state devices are applied for a microwave heating system, like a microwave oven, new microwave heating methods are realized and new applicators can be used with solid state devices. Frequency, phase, and power of the microwaves can be broadly controlled with the solid state devices. It is a merit in microwave chemical science to estimate the effects of the microwave frequency. If frequency and phase of the microwaves in the solid state devices can be controlled, then the power distribution in the applicator and in space can also be controlled. In this chapter, recent research and development (R&D) status of solid state devices and the R&D of RF heating systems that employ solid state devices are reviewed.

1.1 Introduction

Radio frequency (RF) energy can be generated from electricity either via a vacuum tube or a solid state device. RF is very close to radio, audio, and television systems. The only difference between the two applications is that we use weak and modulated RF in radio, audio, and television broadcasts, whereas high power and non-modulated RF is necessary for heating systems such as microwave ovens. Although vacuum tubes were used in the past, solid state devices have also been used recently in radio,

N. Shinohara (✉)
Research Institute for Sustainable Humanosphere (RISH), Kyoto University, Uji 611-0011, Japan
e-mail: Shino@rish.kyoto-u.ac.jp

S. Horikoshi and N. Serpone (eds.), *RF Power Semiconductor Generator Application in Heating and Energy Utilization*, https://doi.org/10.1007/978-981-15-3548-2_1

audio, and television systems. Owing to the required high power for microwave heating systems, vacuum tubes—otherwise also known as cavity magnetrons—are applied. It is easy to generate directly high power RF electrically via the vacuum tube, while it is challenging to generate directly high power RF from a solid state device; usually, an amplifier is incorporated after the low power RF from a solid state device. The development of highly efficient, high power amplifiers (HPA) to be used in conjunction with solid state devices presents a significantly difficult challenge. However, owing to recent advances in solid state device technology, HPAs can be applied in microwave ovens. A wide band gap material such as gallium nitride (GaN) is expected to be a good candidate for high power RF energy applications. GaN is typically applied as a blue light-emitting diode; nonetheless, it can also be employed for power devices (e.g., converters) [1]. Active research and development (R&D) has resulted in many commercial GaN products worldwide [2]. In contrast to wide band gap semiconductor-based devices, historically, conventional silicon (Si)-based solid state devices were considered incompatible for high power applications. Recently, however, Si-based laterally diffused metal–oxide semiconductors (LDMOSs) have been successfully applied in RF energy systems [3]. In this chapter, recent R&D status of solid state devices and the R&D of RF heating systems that use solid state devices are reviewed. The central focus is the technology of RF microwaves, which span a frequency of approximately 1–30 GHz and a wavelength of ca. 30–1 cm.

1.2 Basic Technology of a Microwave Amplifier with a Solid State Device

When discussing solid state devices, it is typical that only amplifiers, not generators, are considered because the key technology of high power microwaves is the amplifier, which is composed of a solid state device and circuits. The solid state device is classified by its material and form, while the amplifier circuit is classified by its level of amplification (Fig. 1.1) [4].

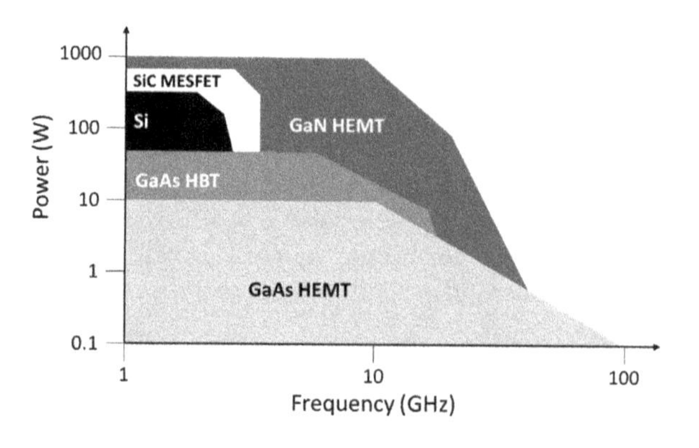

Fig. 1.1 Typical frequency and power relationships of various solid state devices

The solid state device is composed of a p-type semiconductor, n-type semiconductor, and other materials (e.g., metals). Normally, for radio, audio, and television or wireless communication systems, the semiconductor is Si or gallium arsenide (GaAs). However, Si and GaAs cannot generate high power because of material limitations. Consequently, new semiconductor materials are sought for high power microwave applications. GaN and silicon carbide (SiC) are both wide band gap semiconductors that can be used in high power applications. In general, Si and SiC are applied for low frequency applications, and GaAs and GaN are applied for high frequency implementation, i.e., microwaves. Recently, Si characteristics were improved, and some Si-based semiconductor devices can be applied toward high frequency and high power applications. They can be combined to produce devices. In the case of an amplifier, generally a three-terminal device is used, a transistor, which is typically combined with p-n-p or n-p-n semiconductor junctions. The RF signal can be amplified by the transistors. However, a transistor is not suitable at microwave frequencies. The typical solid state device used in microwave amplifier circuits includes field-effect transistors (FETs), heterojunction bipolar transistors, and high electron mobility transistors (HEMTs), all of which have the same three-terminal device structure, but the form and shape are different in the transistor. Recently, metal–oxide semiconductor (MOS) FETs (MOSFETs), which can be applied at higher frequencies, have been applied in microwave systems. In particular, LDMOSs are often applied for high power microwave applications. A combination of the semiconductor materials and the form of the three-terminal devices creates solid state devices such as the Si LDMOS, GaAs FET, and GaN HEMT. The materials' properties determine frequency, power, conversion efficiency, and amplifier gain (Figs. 1.1 and 1.2a).

In order to realize a microwave amplifier, circuit design must be carefully considered. There are three terminals in the three-terminal solid state device. For example, in a FET, the terminals are known as the gate, source, and drain. It is important to decide which terminal is connected to ground and which terminals are signal input and output. Usually, the source is connected to ground, the gate is an input, and the drain is the output in the microwave amplifier circuit. In the RF amplifier circuit, a bias voltage is added to amplify the input RF. Figure 1.2b shows a typical basic amplifier circuit with the solid state device at low frequency. In order to add bias voltage effectively and to amplify the input microwave with high stability, feedback loop circuits are introduced in the basic amplifier circuit. Depending on the bias voltage, the efficiency and gain are determined by the circuit design. Class **A**, **B**, and **C** amplifiers are classified by the bias voltage used in the device and in the same amplifier circuit. These classes can be applied not only to GHz systems but also to kHz–MHz systems. The theoretical RF amplifier efficiency is 50% for class **A**, 78.5% for class **B**, and <100% for class **C**. In classes **B** and **C**, the efficiency is better than in class **A**. However, the waveform is distorted, and wave linearity cannot be maintained. For class **C**, the output power tends to be near zero when the efficiency is close to 100%. As a result, class **B** and **C** devices are not typically used for wireless

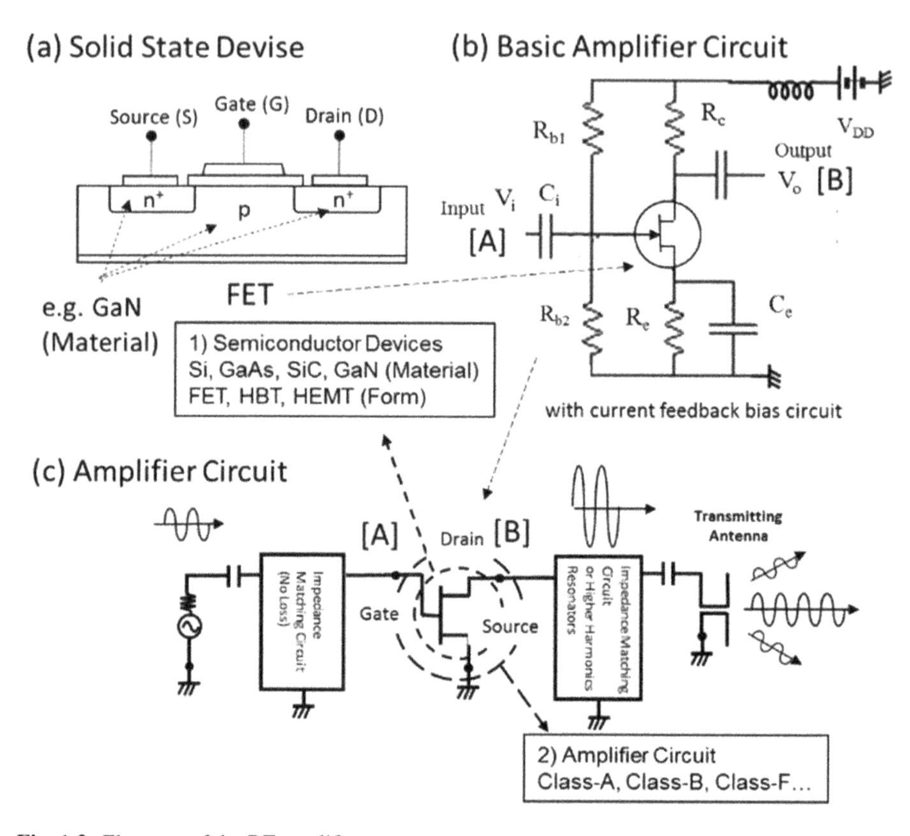

Fig. 1.2 Elements of the RF amplifier

communication systems that require high wave linearity. Before and after the amplifier circuit, impedance matching circuits must be installed to increase the efficiency, the gain, and the power.

Using a basic amplifier circuit is not sufficient to increase the efficiency and obtain higher gain of the amplification. For microwave frequency, instead of the circuit shown in Fig. 1.2b, we must consider impedance matching and the combination of higher frequency to increase the efficiency and gain (Fig. 1.2c). A photograph of a typical microwave amplifier circuit is shown in Fig. 1.3. The circuit is composed of a solid state device and distributed line, which is a thin conductor line on a dielectric substrate known as a microstrip line. When the length of the distributed line changes, the phase of the microwave changes, as well as the impedance that consists of resistance R, inductance L, and capacitance C. When the impedances of two connected circuits are different, microwave reflection occurs, and the efficiency is reduced. As shown in the schematic of Fig. 1.3, distributed lines are installed to match the impedance of two circuits.

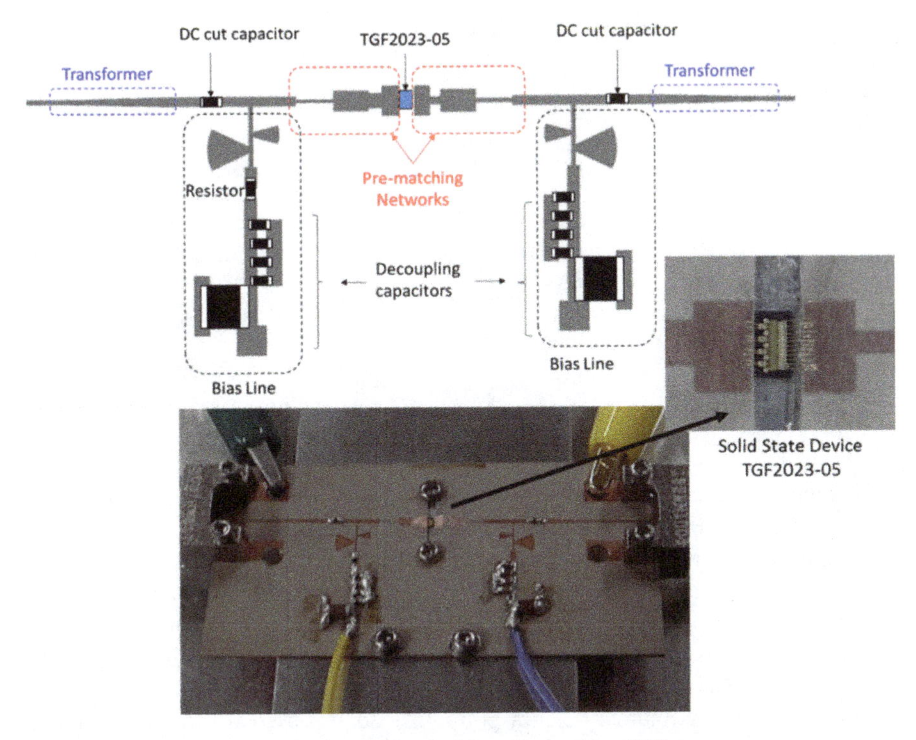

Fig. 1.3 Photograph of a basic microwave amplifier circuit with a FET

In output impedance matching circuit, higher harmonics resonators are applied to increase the efficiency. Class **D** and class **E** amplifiers can be applied not only to kHz–MHz systems but also to GHz systems. Additionally, class **F** and class $\mathbf{F}^{-1}$ amplifiers are often applied to GHz systems, which can theoretically realize 100% efficiency. It is important to increase the number of harmonics to achieve higher efficiency. Some class **F** amplifiers have been developed for a wireless power transfer application [5, 6], which is an RF energy application. The drain efficiency, which is the ratio of output RF power to input direct current (DC) power when the primary input DC power is fed to the drain of the FET, reached 80.1%, and the power added efficiency (PAE = $(P_{\text{out}} - P_{\text{in}})$/PDC) had a maximum of 72.6% at 1.9 GHz in the developed class **F** amplifier using a GaN HEMT [7]. The same research group developed a class **F** amplifier operating at 5.65 GHz using an AlGaN/GaN HEMT. The drain efficiency was 90.7%, the PAE had a maximum of 79.5%, and the saturated power was 33.3 dBm [8]. When combined with higher frequencies, instead of high PAE in class **D**, **E**, **F**, and $\mathbf{F}^{-1}$ operation, the bandwidth of the amplified frequency becomes narrower. This is not suitable for wireless communication applications because they require wide bands for modulated RF. However, for RF energy applications, the bandwidth is typically not required. Therefore, classes **D**, **E**, **F**, and $\mathbf{F}^{-1}$ operation amplifiers are suitable for RF energy applications.

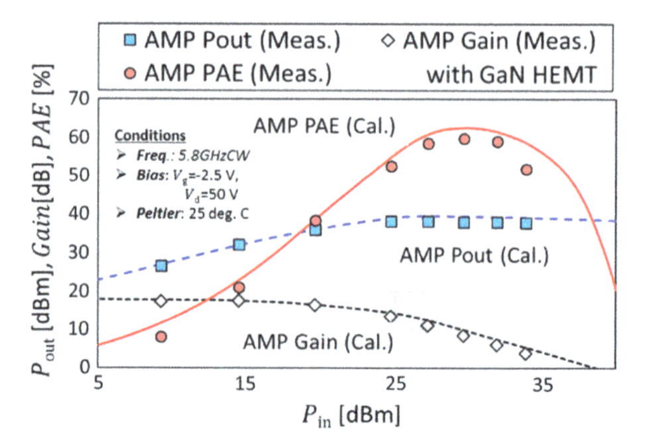

Fig. 1.4 Typical characteristics of a microwave amplifier in terms of P_{out}, gain, and PAE versus input power

In general, the efficiency, gain, and output microwave power depend on input microwave power. Figure 1.4 indicates typical characteristics of the efficiency, gain, and output microwave power versus change of the input microwave power. The PAE is saturated after reaching its maximum. At and after the maximum PAE point, the output microwave is usually distorted, and unexpected harmonics occur. The linearity of the microwave is not good at such a point. When considering the linearity of the microwave, for example, for general wireless communication systems, the amplifier is not used at the maximum efficiency point; instead, it is used at a lower PAE point. Usually, the linearity of the microwave is not focused on RF energy applications. However, for advanced RF energy applications, the linearity should be maintained.

1.3 Recent Research and Development Status of Microwave Amplifiers

In general, when the RF frequency increases, the efficiency, gain, and power of the amplifier decrease. This trend depends mainly on the properties of the solid state device; also, an increase in the parasitic capacitance/inductance by higher frequencies and accuracy/error of the circuit for shorter wavelengths occurs. The output power dependence on the frequency of the developed microwave amplifiers is shown in Fig. 1.5. CW indicates a continuous wave and the pulse is a characteristic of pulse operating amplifiers. For RF energy applications, CW is usually used, and the power is the most important parameter. GaN is expected to be used in high power solid state devices. If the required high power for the RF energy application cannot be generated or developed with one solid state device, semiconductors are combined in one chip, in one amplifier with a power combination, and/or in space after microwave radiation. For example, a power combined microwave amplifier developed by the Japan Aerospace Exploration Agency is illustrated in Fig. 1.5. It consists of 20 amplifiers

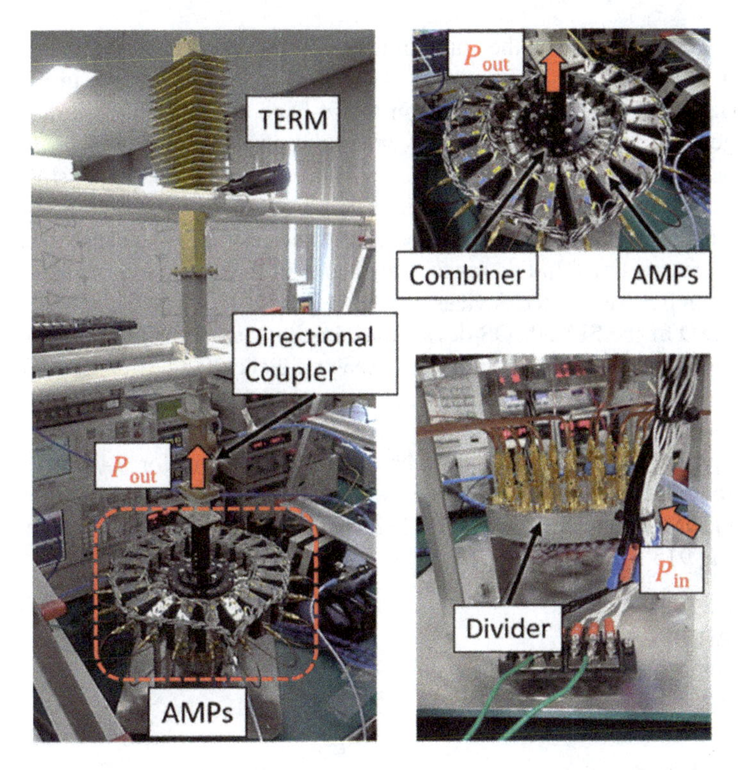

Fig. 1.5 7.1 GHz, 170 W solid state power amplifier with a 20-way combination

and a 20-way power combination. The frequency is 7.1 GHz, and the total output microwave power is 170 W from one output port.

Amplifiers with combined semiconductors on one chip and amplifiers combined with power combinations are shown in Fig. 1.6 [9–16]. In general, the output power and the efficiency decrease when the frequency increases.

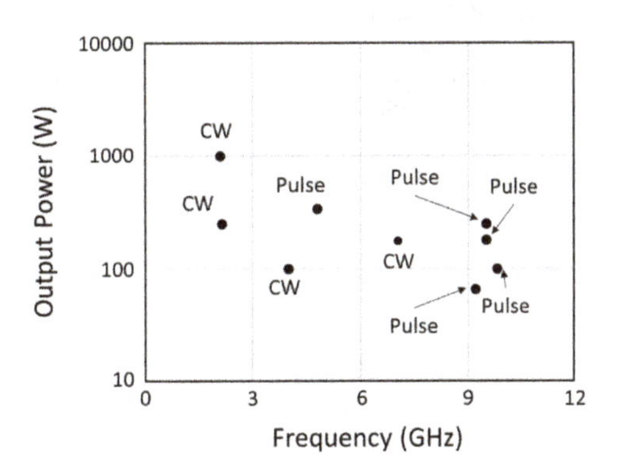

Fig. 1.6 Output power frequency dependence of developed microwave amplifiers—Summarized from ref. [9–16]

The PAE dependence on the output microwave power of the developed amplifiers is shown in Fig. 1.7 in the S-band (2–4 GHz) [17–26], in Fig. 1.8 in the C-band (4–8 GHz) [27–38], and in Fig. 1.9 in the X-band (8–12 GHz) [12, 14–16, 39–51]. In general, at the same frequency when the power of the amplifier increases, the efficiency decreases. If the PAE is low, an extra heat reduction system may be necessary because the solid state device cannot be operated at high temperatures. The efficiency will decrease at higher temperatures, and heat will cause the device to eventually break down. In this sense, GaN is one of the hopeful candidate materials to be used in heat resistance devices.

The R&D of the Si LDMOS device has already been applied in commercial RF energy applications. The evolution of the peak efficiency for a 30-V LDMOS device is presented in Fig. 1.10. The efficiency increases to 67% at 2.14 GHz for a 150 W device and to 55% at 3.6 GHz for a 10-W device [3]. In contrast with the GaN technology, the LDMOS technology has become sufficient for S-band RF energy applications. Recently commercialized were a 1 kW microwave amplifier module using an LDMOS device at 2.45 GHz [52] and a 64 kW RF heating system with an LDMOS at 915 MHz bands [53].

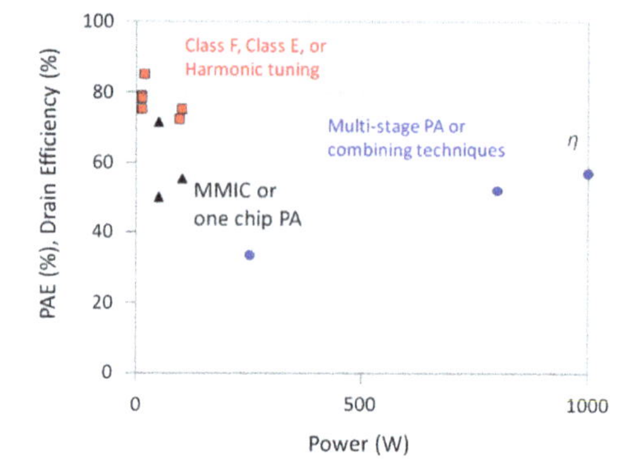

Fig. 1.7 PAE dependence on the microwave output power of developed microwave amplifiers in the S-band (2–4 GHz)—Summarized from ref. [17–26]

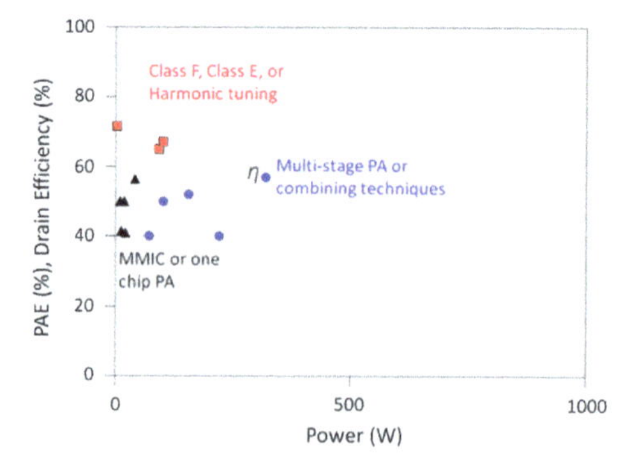

Fig. 1.8 PAE dependence on the microwave output power of developed microwave amplifiers in the C-band (4–8 GHz)—Summarized from ref. [27–38]

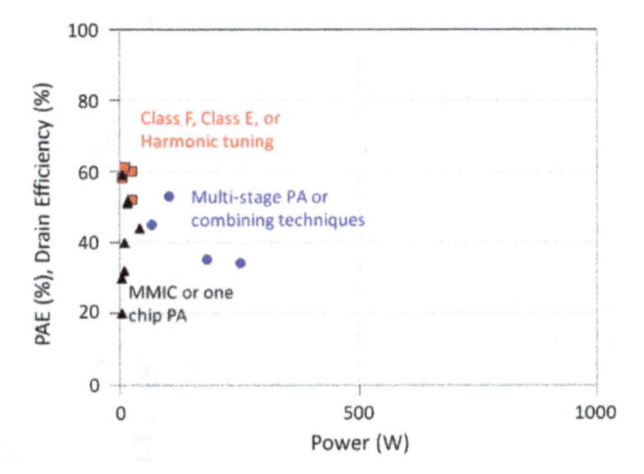

Fig. 1.9 PAE dependence on the microwave output power of developed microwave amplifiers in the X-band (8–12 GHz)—Summarized from ref. [12, 14–16, 39–51]

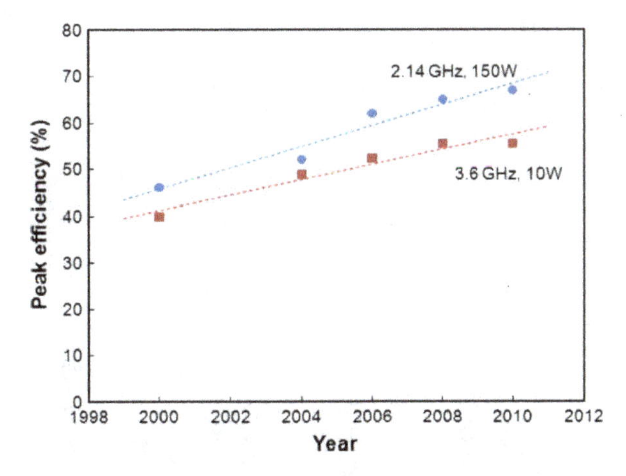

Fig. 1.10 Evolution of the peak efficiency for a 30 V LDMOS device over time. The efficiency increases to 67% at 2.14 GHz for a 150 W device and to 55% at 3.6 GHz for a 10 W device. Lines are a guide for the eye—Reproduced from ref. [3]. Copyright by IEEE

1.4 Recent Commercial High Power Microwave Amplifiers

It has become easy to find commercial solid state devices and amplifiers for RF energy applications. Currently, commercial semiconductors and modules/systems are mainly based on LDMOS technology in the 915 MHz and 2.45 GHz bands; indeed some companies produce and develop GaN devices and amplifier systems for use in higher frequency bands (see Tables 1.1 [54–64] and 1.2 [65–78]). In the near future, GaN devices and related modules/systems are also likely to be commercial products.

Table 1.1 Commercial semiconductor products for RF energy applications

Company	Device	Frequency	Power (CW) (W)	Gain (dB)	Drain Eff.	Semiconductor	References
Ampleon	BLF0910H9LS750P	915 MHz	750	21.5	72.5%	LDMOS	[54]
	BLC2425M10LS500P	2.45 GHz	500	15.0	67.5%	LDMOS	[55]
Innogration technologies	ITDE10700D4	915 MHz	700	15.0	68%	LDMOS	[56]
	ITCH25280D4	2.45 GHz	300	13.5	57%	LDMOS	[57]
	GTBV25500D4	2.45 GHz	500	15.0	68%	GaN (develop)	[58]
	GTBV58080G4	5.8 GHz	80	12.0	50%	GaN (develop)	[59]
NXP	MRF24300N	2.45 GHz	320	13.1	60.5%	LDMOS	[59]
Sumitomo electric device innovations	SGNE090MK	900 MHz	125	20	70%	GaN	[60]
	EGN26C210I2D	2.6 GHz	200	16	62%	GaN	[60]
	SGK5867-100A/001	5.85–6.75 GHz	112	13.5	45% (PAE)	GaN	[61]
	SGK1314-60A	13.75–14.5 GHz	63	8.5	32% (PAE)	GaN	[62]
MACOM	MAGe-102425-300	2.45 GHz	300	15.0	70%	GaN	[63]
Mitsubishi electric	MGFK50G3745	13.75–14.5 GHz	100	9.2	30% (PAE)	GaN	[64]

Table 1.2 Commercial modules and amplifier systems for RF energy applications

Company	Module/System	Frequency	Power (CW)	Efficiency	Semiconductor	References
SAIREM	System	915 MHz	600 W	?	?	[65]
	System	2.45 GHz	450 W	?	?	[66]
Ampleon	Module	2.4–2.5 GHz	250 W	?	LDMOS	[67]
	System	2.45 GHz	1 kW	50%	LDMOS	[68]
NXP	Module	2.45 GHz	250 W	?	LDMOS	[69]
Chengdu Wattsine electronic technology	System	915 MHz	20 kW	40%	LDMOS	[70, 71]
	System	2.45 GHz	20 kW	40%	LDMOS	[70, 71]
Fricke und Mallah microwave technology	System	2.45 GHz	500 W	60%	?	[72]
Tokyo Keiki	System (develop)	2.45 GHz	100 W	60%	GaN	[73]
	System	9–10 GHz (pulse)	100 W	36%	GaN	[74]
Mitsubishi electric	System (develop)	2.45 GHz	500 W	?	GaN	[75]
R&K	System	80 MHz–1 GHz	1.5 kW	Class A	?	[76]
	System	2–4 GHz	500 W	?	?	[77]
	System	2.5–6 GHz	250 W	?	?	[78]

1.5 New Microwave Heating Systems with Solid State Devices

RF energy is usually applied in microwave heating, e.g., in a microwave oven. Based on this technology, a new scientific field has emerged now known as microwave chemical science. In microwave chemical science, an applicator is used, which is a microwave shielded box. Typical applicators are shown in Fig. 1.11. There are two types of applicators, single-mode applicators and multi-mode applicators. A single-mode applicator uses a traveling wave from a microwave source, a reflected wave

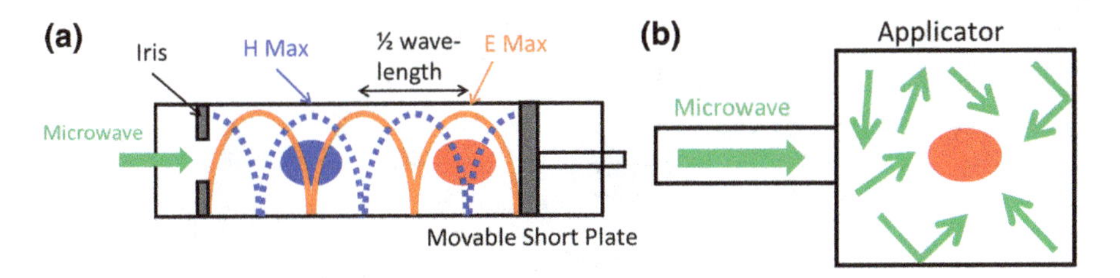

Fig. 1.11 **a** Single-mode and **b** multi-mode microwave applicators

from a short (reflect) plane, and a standing wave as the interference of two waves. As a result of the presence of a standing wave, the electric and magnetic fields of the electromagnetic waves (microwaves) can be separated. In the single-mode applicator, only the electric field or magnetic field is used to heat materials. The multi-mode applicator is like a microwave oven. The component fields of the electromagnetic wave cannot be separated, but this type of system readily heats materials.

When solid state devices are applied for a microwave heating system such as a microwave oven, new microwave heating methods are realized, and new applicators can be used with solid state devices. Frequency, phase, and power of the microwaves can be broadly controlled with the solid state devices. It is a merit of microwave chemical science to be able to estimate the effects of the microwave frequency. With a magnetron, it is difficult to control the frequency because resonators, which have narrow band characteristics, are adopted in the magnetron to generate microwaves at a stable frequency. The low power microwave generation in the magnetron is unstable, and the microwave power cannot be decreased with the magnetron.

When the microwave frequency is changed, a suitable applicator for various frequencies should be considered. As shown in Fig. 1.11, the single-mode applicator is a waveguide with a cut off frequency. The microwave propagates through the waveguide in the multi-mode applicator. The microwave with a lower frequency than the cutoff frequency cannot travel in the waveguide. Therefore, instead of a waveguide, a microwave irradiation probe, based on a coaxial line and a cylindrical applicator, is proposed (see Fig. 1.12) [79]. There is no cutoff frequency in the coaxial line when the propagating microwave mode is in a transverse electromagnetic mode. Figure 1.13 shows a photograph of the developed cylindrical applicator with the microwave irradiation probe [79]. The optical fiber thermometer port was attached orthogonally to the microwave input port on the side of the applicator. The pressure gage was mounted to the top of the applicator to maintain the internal pressure. Simulated and measured values of the reflection ratios for the developed cylindrical applicator using liquid samples (water and NaOH solution) are shown in Fig. 1.14a, [79]. Figure 1.14b displays the measured temperature increase in water and in a NaOH solution during microwave heating. Wide frequency microwaves propagate

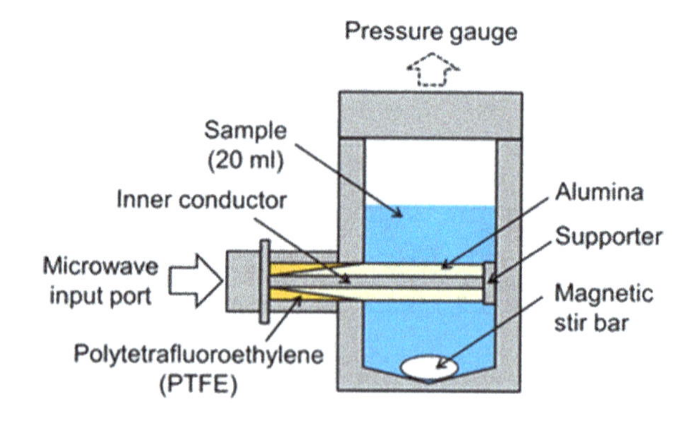

Fig. 1.12 Cross-sectional schematic diagram of the proposed microwave irradiation probe and cylindrical applicator—Reproduced from ref. [79] Copyright Processes by MDPI

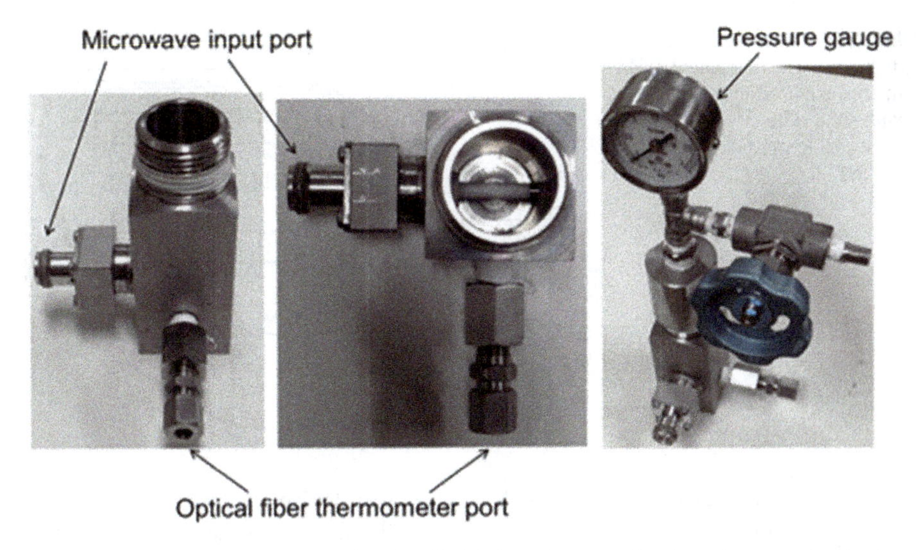

Fig. 1.13 Photographs of the developed cylindrical applicator with the microwave irradiation probe—Reproduced from ref. [79]. Copyright processes by MDPI

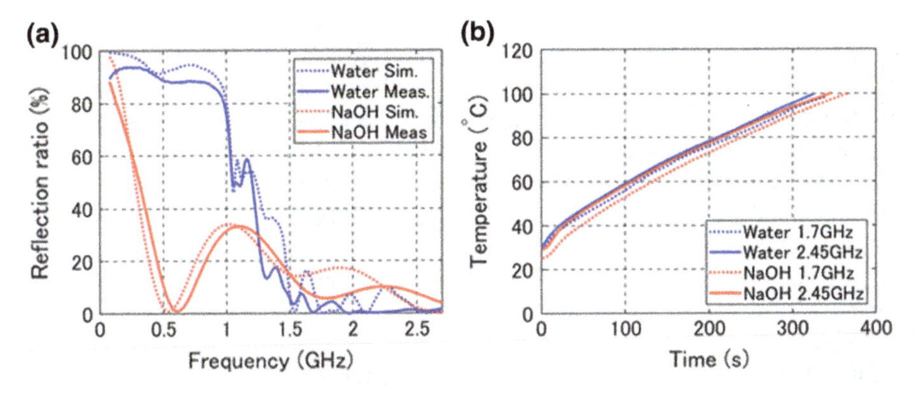

Fig. 1.14 **a** Simulated and measured results of the reflection ratios for the developed cylindrical applicator using liquid samples (water and an NaOH solution) and **b** measured temperature increase in water and an NaOH solution during microwave heating at different frequencies—Reproduced from ref. [79]. Copyright processes by MDPI

into the cylindrical applicator, and the liquids are well heated with multi-frequency radiation.

If control frequency and phase of the microwave in the solid state devices can be controlled, then the power distribution in the applicator and in space can be controlled. Solid state devices are used in one applicator. In the applicator, the radiated microwaves are combined in space. This technology is based on radar systems and is called a phased array antenna that can control beam direction by combining the controlled phases of each antenna. The phases cannot be controlled; it is sufficient

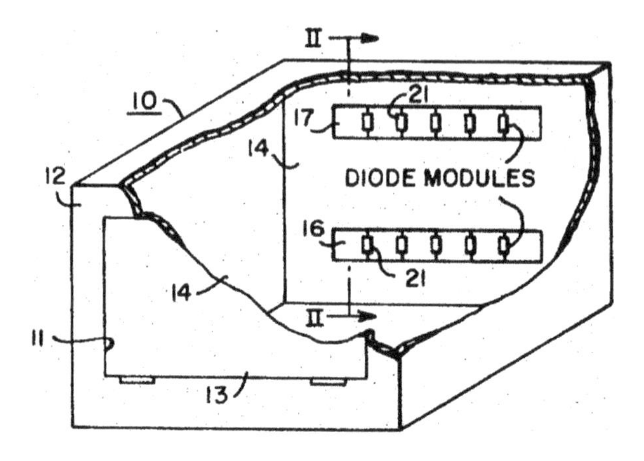

Fig. 1.15 Schematic diagram of the first patented microwave heating system with solid state devices—Reproduced from ref. [81]

to increase the microwave power with small power devices. When comparing, the microwave power from a magnetron is still smaller than that of solid state devices. Nonetheless, power combination in space is necessary.

The idea of a microwave heating system with solid state devices was proposed in the late 1960s [80]. The first patent of the microwave oven with solid state devices was granted in 1971 (Fig. 1.15) [81]. In the patent, instead of Si LDMOS or GaN HEMT, a diode microwave generator was adopted. The diode is a two-terminal device. The diode can generate microwaves, but it cannot change the phase because it is a generator. On the other hand, there are many diodes in one applicator, making it similar to other solid state devices. In the 1980s and 1990s, some research groups were granted patents for a microwave oven in which the frequency was controlled for an optimal spatial pattern (Fig. 1.16) [82–84]. In Fig. 1.17, the effect of spatial pattern changes by frequency control from a two-port applicator (shown in Fig. 1.18) is presented [85]. Spatial pattern of the electric field can be controlled by frequency control.

In parallel to the development of the phased array antenna for radar systems, a solid state microwave generating array system for which each element is phase controllable was proposed and granted in 1995 (Fig. 1.19) [86]. When comparing the frequency control with the special pattern control, the phase control readily controls the special pattern. In free space without a wall, it is easy to estimate the microwave focusing point by the phase-shifting technology. However, in the applicator, there are walls, and it is a multi-pass (multi-reflection) circumstance. Therefore, considering the multi-reflection circumstance to focus on the expected point in the applicator is a challenging task. Extra phase shifters are required in each solid state device, and the phase shifter is expensive. Recently, for mobile communication systems, a digital beam-forming network system with digital phase shifters was installed; spinoff technology from the mobile communication system is expected in the near future. A new microwave oven with controllable GaN amplifiers has been developed at Sophia University (see Fig. 1.20) [87, 88]. The microwave power is controlled on an expected focal point only.

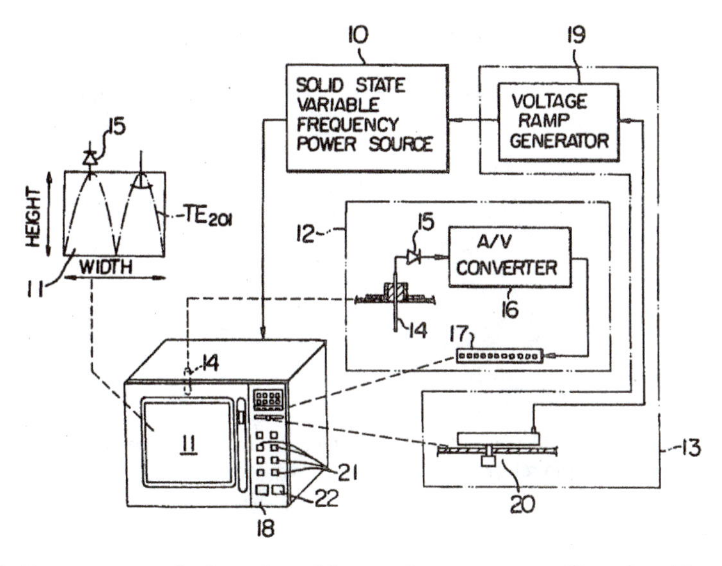

Fig. 1.16 Frequency control scheme for solid state microwave ovens—Reproduced from ref. [84]

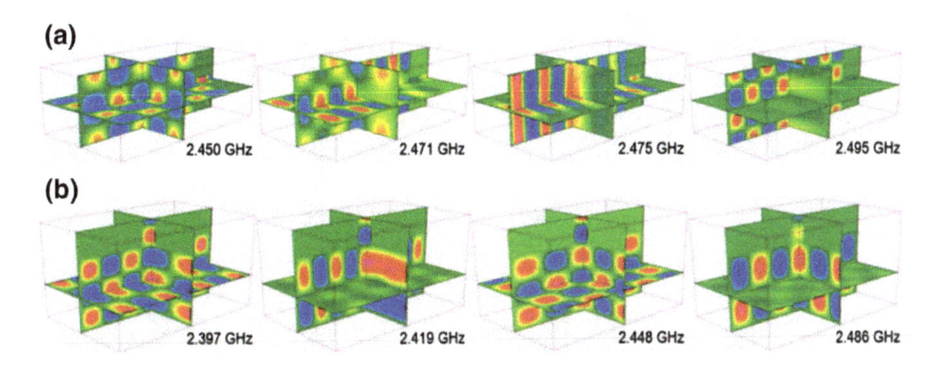

Fig. 1.17 Patterns of an instantaneous electric field visualized in coordinate planes at resonant frequencies in an empty system (shown in Fig. 1.18) excited by ports I (**a**) and II (**b**) in the 2.4–2.5 GHz frequency range—Reproduced from ref. [85] Copyright by AMPERE

1.6 Concluding Remarks

Recently, solid state devices have been implemented in RF energy applications because frequency and phase can be controlled to modify the power distribution; both frequency and power can be optimized to consider the frequency effects in microwave chemistry. Si LDMOSs are already applied in commercial low microwave frequency RF energy applications. The GaN HEMT is expected to enable the development of high power and higher microwave frequency. Current requirements in the field

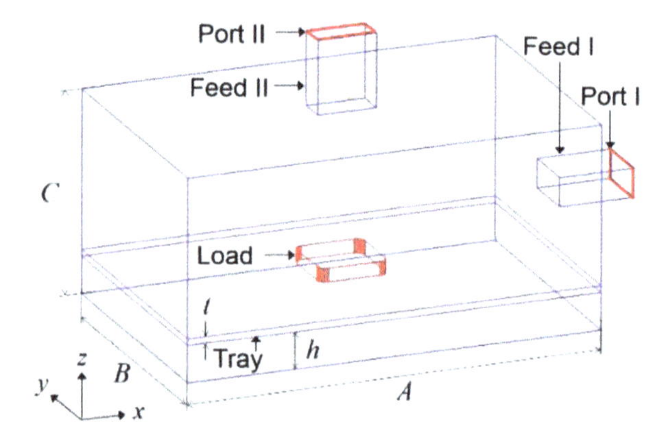

Fig. 1.18 Three-dimensional view of loaded rectangular cavity for the simulation shown in Fig. 1.17—Reproduced from ref. [85]. Copyright by AMPERE

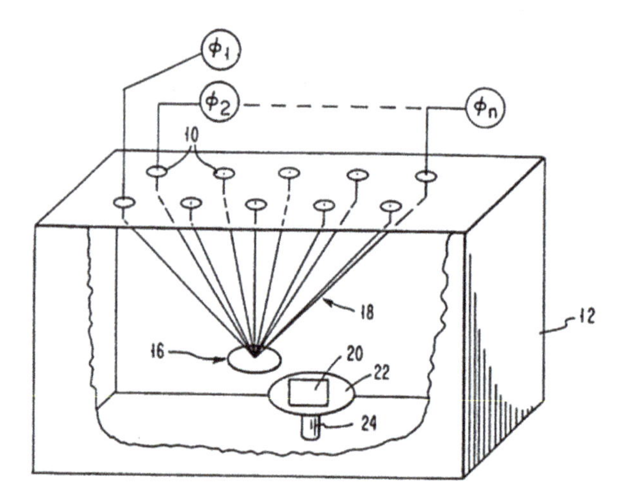

Fig. 1.19 Solid state microwave generating array system; each element is phase controllable—Reproduced from ref. [86]

of microwave chemistry provide new solid state device applicators. However, solid state device technology is not just used in RF energy applications. It is also applied in mobile communication and radar technologies. In the near future, new solid state devices may be used to meet the demands of society.

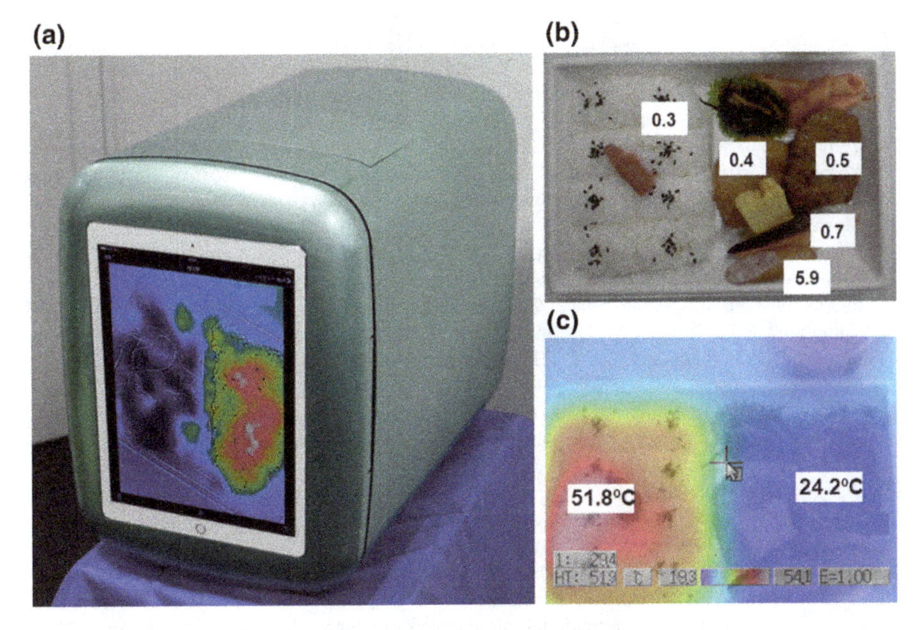

Fig. 1.20 **a** Intelligent microwave oven with phase controllable GaN amplifiers [87]; **b** tan(δ) of topping food included in a lunch box and **c** the temperature distribution in the lunch box during microwave selective heating—Reproduced from ref. [88]. Copyright by Sophia University and IMPI

References

1. Jones EA, Wang F, Costinett D (2016) Review of commercial GaN power devices and GaN-based converter design challenges. IEEE J Emerg Sel Top Power Electron 4(3):707–719
2. Mudassir S, Muhammad J (2013) A review of gallium nitride (GaN) based devices for high power and high frequency applications. J Appl Emerg Sci 4(2):141–146
3. Theeuwen SJCH, Qureshi JH (2012) LDMOS technology for RF power amplifiers. IEEE Trans Microw Theory Tech 60(6):1755–1763
4. Oliver S (2014) Optimize a power scheme for these transient times. Electron Des
5. Shinohara N (2014) Wireless power transfer via radiowaves (wave series). ISTE Ltd. and Wiley, London, UK and Hoboken, USA
6. Shinohara N (ed) (2018) Recent wireless power transfer technologies via radio waves. River Publishers, Delft, Netherlands
7. Zheng C, Yoshida T, Ishikawa R, Honjo K (2007) GaN HEMT class-f amplifier operating at 1.9 GHz (*in Japanese*). In: Proceedings of IEICE, Nagoya, Japan, C-2-27, 20–23 Mar 2007
8. Kamiyama M, Ishikawa R, Honjo K (2011) C-band high efficiency AlGaN/GaN HEMT power amplifier by controlling phase angle of harmonics (in Japanese). In: Proceedings of IEICE, Sapporo, Japan, CS-3-1, 13–16 Sep 2011
9. Hasegawa N, Shinohara N, Kawasaki S (2016) A 7.1 GHz 170 W solid-state power amplifier with 20-way combiner for space applications. IEICE Trans Electron 99-C(10):1140–1146
10. Kobayashi Y, Yoshida Y, Yamamoto Z, Kawasaki S (2013) S-band GaN on Si based 1 kW-class SSPA system for space wireless applications. IEICE Trans Elec E96-C(10): 1245–1253
11. Takenaka I, Ishikawa K, Asano K, Takahashi S, Murase Y, Ando Y, Takahashi H, Sasaoka C (2014) High-Efficiency and high-power microwave amplifier using GaN-on-Si FET with improved high-temperature operation characteristics. IEEE Trans Microw Theory Thech 62(3):502–512

12. Shigematsu H, Inoue Y, Akasegawa A, Yamada M, Masuda S, Kamada Y, Yamada A, Kanamura M, Ohki T (2009) C-band 340-W and X-band 100-W GaN power amplifiers with over 50-% PAE. In: Proceedings of IEEE international microwave symposium digest, Boston, USA, 7–12, June 2009, pp 1265–1268
13. Yamasaki T, Kittaka Y, Minamide H, Yamauch K, Miwa S, Goto S, Nakayama M, Kono M, Yoshida N (2010) A 68% efficiency, C-band 100 W GaN power amplifier for space applications. In: Proceedings of IEEE MTT-S international microwave symposium, Anaheim, USA, 23–28 May 2010, pp 1384–1387
14. Casto M, Lampenfeld M, Jia P, Courtney P, Behan S, Daughenbaugh P, Worley R (2011) 100 W X-band GaN SSPA for medium power TWTA replacement. In: Proceedings of IEEE wireless microwave technology conference, Clearwater Beach, USA, 18–19 Apr 2011, pp 1–4
15. Campbell CF, Poulton M (2011) Compact highly integrated Xband power amplifier using commercially available discrete GaN FETs. In: Proceedings of Asia-Pacific microwave conference, Melbourne, Australia, 5–18 Dec 2011, pp 243–246
16. Kanto K, Satomi A, Asahi Y, Kashiwabara Y, Matsushita K, Takagi K (2008) An X-band 250 W solid-state power amplifier using GaN power HEMTs. In: Proceedings of IEEE Radio Wireless Symposium, Orlando, USA, 22–24 Jan 2008, pp 77–80
17. Wang Y, Dong S, Yang L, Li Z, Dong Y, Fu W (2014) Design of high efficiency GaN HEMT class-F power amplifier at S-band. In: Proceedings of Asia-Pacific conference antennas and propagation, Harbin, China, 26-29 July 2014
18. Motoi K, Matsunaga K, Yamanouchi S, Kunihiro K, Fukaishi M (2012) A 72% PAE, 95-W, single-chip GaN FET S-band inverse class-F power amplifier with a harmonic resonant circuit. In: Proceedings of IEEE international microwave symposium, Montreal, Canada, 17–22 June 2012, pp 1–3
19. Saad P, Nemati HM, Thorsell M, Andersson K, Fager C (2009) An inverse class-F GaN HEMT power amplifier with 78% PAE at 3.5 GHz. In: Proceedings of European microwave conference, Rome, Italy, 30 Sep–2 Oct 2009, pp 496–499
20. Schmelzer D, Long SI (2007) A GaN HEMT class F amplifier at 2 GHz with >80% PAE. IEEE J Solid-State Circuits 42(10):2130–2136
21. Ui N, Sano S (2006) A 100 W Class-E GaN HEMT with 75% drain efficiency at 2 GHz. In: Proceedings of European microwave integrated circuits conference, Manchester, UK, pp 10–13
22. Kim J, Moon J, Kim J, Boumaiza S, Kim B (2009) A novel design method of highly efficient saturated power amplifier based on self-generated harmonic currents. In: Proceedings European microwave conference, Rome, Italy, 30 Sep–2 Oct 2009, pp 1082–1085
23. High power, high efficiency, AlGaN/GaN HEMT technology for wireless base station applications. In: Proceedings of IEEE international microwave symposium, Long Beach, USA, 12–17 June 2005
24. Kikkawa T, Nagahara M, Adachi N, Yokokawa S, Kato S, Yokoyama M, Kanamura M, Yamaguchi Y, Hara N, Joshin K (2003) High-power and high-efficiency AlGaN/GaN HEMT operated at 50 V drain bias voltage. In: Proceedings of IEEE radio frequency integrated circuits (RFIC) symposium, Philadelphia, USA, 8–13 June 2003, pp 167–170
25. Iqbal M, Piacibello A (2016) A 5W class-AB power amplifier based on a GaN HEMT for LTE communication band. In: Proceedings of 16th mediterranean microwave symposium, Abu Dhabi City, UAE, 14–16 Nov 2016, pp 1–4
26. Mitani E, Aojima M, Sano S (2007) A kW-class AlGaN/GaN HEMT pallet amplifier for S-band high power application. In: Procedings of European microwave integrated circuit conference, Munich, Germany, 8–12 Oct 2007, pp 176–179
27. Giordani R, Amici RM, Barigelli A, Conti F, Del Marro M, Feudale M, Imparato M, Suriani A (2009) Highly integrated and solderless LTCC based-band T/R module. In: Proceedings of European microwave conference, Rome, Italy, 28 Sep–2 Oct 2009, pp 1760–1763
28. de Hek AP, de Boer A, Svensson T (2001) C-band 10-watt HBT high-power amplifier with 50% PAE. In: Proceedings of gallium arsenide application symposium, London, UK, 24–28 Sep 2001, pp 1–3

29. Lhortolary J, Ouarch Z, Chang C, Camiade M (2009) A 17W C-band high efficiency high power pHEMT amplifier for space applications. In: Proceedings of European microwave integrated circuits conference, Rome, Italy, 28 Sep–2 Oct 2009, p 21
30. Florian C, Cignani R, Niessen D, Santarelli A (2012) A C-band AlGaN-GaN MMIC HPA for SAR. IEEE Microw Wirel Compon Lett 22(9):471–473
31. Kido M, Kawasaki S, Shibuya A, Yamada K, Ogasawara T, Suzuki T, Tamura S, Seino K, Ichikawa A, Tsuchiko A (2016) 100 W C-band GaN solid state power amplifier with 50% PAE for satellite use. In: Proceedings of Asia-Pacific microwave conference (APMC), New Delhi, India, 5–9, Dec 2016
32. Tang H, Wang Z, Xu T, Wang X (2016) High efficiency GaN power amplifier on C band. In: Proceedings of 17th international conference electronic packaging technology, Wuhan, China, 16–19 Aug. 2016, pp 1413–1417
33. Hirano T, Shibuya, A, Kawabata T, Kido M, Yamada K, Seino K, Ichikawa A, Kamikokura A (2014) 70 W C-band GaN solid state power amplifier for satellite use. In: Proceedings of Asia-Pacific microwave conference (APMC), Sendai, Japan, 4–7 Nov 2014, pp 783–785
34. Lu Y, Cao M, Wei J, Zhao B, Ma X, Hao Y (2014) 71% PAE C-band GaN power amplifier using harmonic tuning technology. Electron Lett 50(17):1207–1209
35. Miwa S, Kamo Y, Kittaka Y, Yamasaki T, Tsukahara Y, Tanii T, Kohno M, Goto S, Shima A (2011) A 67% PAE, 100 W GaN power amplifier with on-chip harmonic tuning circuits for C-band space applications. In: Proceedings of IEEE MTT-S international microwave symposium, Baltimore, USA, 5–11 June 2011, pp 1–4
36. Kuroda K, Ishikawa R, Honjo K (2010) Parasitic compensation design technique for a C-Band GaN HEMT class-F amplifier. IEEE Trans Microw Theory Tech 58(11):2741–2750
37. Shigematsu H, Inoue Y, Masuda S, Yamada M, Kanamura M, Ohki T, Makiyama K, Okamoto N, Imanishi K, Kikkawa T, Joshin K, Hara N (2008) C-band GaN-HEMT power amplifier with over 300-W output power and over 50-% Efficiency. In: Proceedings of IEEE compound semiconductor integrated circuits symposium, Monterey, USA, 12–15 Oct 2008, pp 1–4
38. Yamanaka K, Mori K, Iyomasa K, Ohtsuka H, Noto H, Nakayama M, Kamo Y, Isota Y (2007) C-band GaN HEMT power amplifier with 220 W output power. In: Proceedings of IEEE International microwave symposium, Honolulu, USA, 3–9 June 2007, pp 1251–1254
39. Stameroff AN, Ta HH, Pham A, Leoni RE III (2013) Wide-bandwidth power-combining and inverse class-F GaN power amplifier at X-Band. IEEE Trans Microw Theory Tech 61(3):1291–1300
40. Uchida H, Noto H, Yamanaka K, Nakayama M, Hirano Y (2012) An X-band internally-matched GaN HEMT amplifier with compact quasilumped-element harmonic-terminating network. In: Proceedings of IEEE MTT-S international microwave symposium, Montreal, Canada, 17–22 June 2012, pp 1–3
41. Moon JS, Moyer H, Macdonald P, Wong D, Antcliffe M, Hu M, Willadsen P, Hashimoto P, McGuire C, Micovic M, Wetzel M, Chow D (2012) High efficiency X-band class-E GaN MMIC high-power amplifiers. In: Proceedings of IEEE RF power amplifiers for wireless and radio applications topical conference, Santa Clara, USA, 15–18 Jan 2012, pp 9–12
42. Kang J, Moon JS (2017) Highly efficient wideband X-band MMIC class-F power amplifier with cascode FP GaN HEMT. Electronics Lett. 53(17):1207–1209
43. Camarchia V, Fang J, Ghione G, Rubio JM, Pirola M, Quaglia R (2012) X-band wideband 5 W GaN MMIC power amplifier with large-signal gain equalization. In: Proceedings of integrated nonlinear microwave and millimetre wave circuits workshop, Dublin, Ireland, 3–4 Sep 2012, pp 1–3
44. Ersoy E, Meliani C, Chevtchenko S, Kurpas P, Matalla M, Heinrich W (2012) A high-gain -band GaN-MMIC power amplifier. In: Proceedings of 7th German microwave conference, Ilmenau, Germany, 12–14 Mar 2012, pp 1–4
45. Masuda S, Yamada M, Kamada Y, Ohki T, Makiyama K, Okamoto N, Imanishi K, Kikkawa T, Shigematsu H (2012) GaN single-chip transceiver frontend MMIC for -band applications. In: Proceedings of IEEE MTT-S international microwave symposium, Montreal, Canada, 17–22 Jun 2012, pp 1–3

46. Paidi V, Shouxuan X, Coffie R, Moran B, Heikman S, Keller S (2003) High linearity and high efficiency of class-B power amplifiers in GaN HEMT technology. IEEE Trans Microw Theory Tech 51:643–652
47. Shouxuan X, Paidi V, Coffie R, Keller S, Heikman S, Moran B (2003) High-linearity class B power amplifiers in GaN HEMT technology. IEEE Microw Wireless Components Lett 13:284–286
48. Waltereit P, Kuhn J, Quay R, van Raay F, Dammann M, Casar M (2012) High efficiency X-band AlGaN/GaN MMICs for space applications with lifetimes above 10^5 hours. In: Proceedings of 7th european microwave integrated circuit conference (EuMIC), Amsterdam, Netherlands, 29–30 Oct 2012, pp 123–126
49. Shin D, Yom I, Kim D (2017) X-band GaN MMIC power amplifier for the SSPA of a SAR system. In: Proceedings of IEEE international symposium radio-frequency integration technology (RFIT), Seoul, Korea, 30 Aug–1 Sep 2017, pp 93–95
50. Masuda S, Yamada M, Kamada Y, Ohki T, Makiyama K, Okamoto N, Imanishi K, Kikkawa T, Shigematsu H (2011) GaN singlechip transceiver frontend MMIC for X-band applications. In: Proceedings of IEEE international microwave symposium, Baltimore. USA, 5–11 June 2011, pp 1–3
51. Nishihara M, Yamamoto T, Mizuno S, Sano S, Hasegawa Y (2011) X-band 200 W AlGaN/GaN HEMT for high power application. In: Proceedings of 6th European microwave integrated circuit conference, Manchester, UK, 10–11 Oct 2011, pp 65–68
52. Williams R, Lindseth B (2016) Compact 1 kW 2.45 GHz solid-state source for in industrial applications. In: Proceedings of the 50th annual microwave power symposium, Orlando, USA, 21–23 June 2016, pp 39–41
53. Bartola B, Kaplan K, Williams R (2016) 64 kW microwave generator using LDMOS power amplifiers for industrial heating applications. In: Proceedings of the 50th annual microwave power symposium, Orlando, USA, 21–23 June 2016, pp 37–38
54. https://www.ampleon.com/products/rf-energy/915-mhz-transistors/BLF0910H9LS750P.html. Accessed May 2019
55. https://www.ampleon.com/products/rf-energy/2.45-ghz-transistors/BLC2425M10LS500P.html. Accessed May 2019
56. http://www.innogration.net/product/?keys=471. Accessed May 2019
57. http://www.innogration.net/product/?keys=469. Accessed May 2019
58. https://www.nxp.com/docs/en/data-sheet/MRF24300N.pdf. Accessed May 2019
59. https://www.sedi.co.jp/data.jsp?version=&database=wireless&id=6390&class=01010500. Accessed May 2019
60. https://www.sedi.co.jp/data.jsp?version=&database=wireless&id=6582&class=01010202. Accessed May 2019
61. https://www.sedi.co.jp/data.jsp?version=&database=wireless&id=6684&class=01010100. Accessed May 2019
62. https://www.sedi.co.jp/data.jsp?version=&database=wireless&id=6683&class=01010100. Accessed May 2019
63. https://www.macom.com/products/product-detail/MAGe-102425–300. Accessed May 2019
64. http://www.mitsubishielectric.co.jp/semiconductors/content/product/highfrequency/gan/internally/mgfk50g3745.pdf. Accessed May 2019
65. https://www.sairem.com/wp-content/uploads/2017/10/GLS-600-W-EN.pdf. Accessed May 2019
66. https://www.sairem.com/wp-content/uploads/2017/10/GMS-450W_EN.pdf. Accessed May 2019
67. https://www.ampleon.com/products/rf-energy/pallets-and-modules/BPC2425M9X2S250-1.html. Accessed May 2019
68. https://www.rfmw.com/datasheets/ampleon/M2A.pdf. Accessed May 2019
69. https://www.nxp.com/jp/products/rf/rf-power/rf-cooking/2450-mhz-subsystem-for-rf-cooking/2.45-ghz-rf-energy-module:RFEM24-250. Accessed May 2019

70. http://www.wattsine.com/rf-power-amplifier/Solid_state_microwave_source/33.html. Accessed May 2019
71. http://www.wattsine.com/power-source/Medical_equipment_application_solutions/38.html. Accessed May 2019
72. https://www.microwaveheating.net/files/Microwaveheating/Dokumente/Datasheet%20SSMWG%20500W%202450%20Hz_2019-0021_190415.pdf. Accessed May 2019
73. https://www.tokyokeiki.jp/products/detail.html?pdid=214. in Japanese, Accessed May 2019
74. https://www.tokyokeiki.jp/e/products/detail.html?pdid=188. Accessed May 2019
75. https://www.titech.ac.jp/english/news/2016/033205.html. Accessed May 2019
76. http://www.rk-microwave.com/products/pdf/A080M102-6262R.pdf. Accessed May 2019
77. http://www.rk-microwave.com/jp/products/PA.php#prd02. Accessed May 2019
78. http://www.rk-microwave.com/products/pdf/GA252M602-5454R.pdf. Accessed May 2019
79. Mitani T, Nakajima R, Shinohara N, Nozaki Y, Chikata T, Watanabe T (2019) Development of a microwave irradiation probe for a cylindrical applicator. Proceses 7(143)
80. Schwartz E (2016) Historical notes on solid-state microwave heating. APMERE Newsletter 89:4–7
81. McAvoy BR (1971) Solid state microwave oven. US Patent 4097708, 21 Jan 1971
82. MacKay AB (1980) Controlled heating microwave ovens. US Patent 4196332, 1 Apr 1980
83. MacKay AB (1980) Controlled heating microwave ovens using different operating frequencies. CA Patent 1081796, 15 Jul 1980
84. Nobue T, Kusunoki S (1983) Microwave oven having controllable frequency microwave power source. US Patent 4415789, 15 Nov 1983
85. Yakovlev VV (2016) Computer modeling in the development of mechanisms of control over microwave heating in solid-state energy systems. APMERE Newsletter 89:18–21
86. Cuomo JJ, Guarnieri CR, Whitehair SJ (1995) Solid state microwave generating array material, each element of which is phase controllable, and plasma processing systems. EU Patent EP0459177B1, 20 Dec 1995
87. http://rscdb.cc.sophia.ac.jp/seeds/1720_E.html. Accessed May 2019
88. Horikoshi S (2015) Selective heating of food using a semiconductor phase control microwave cooking oven. In: Proceedings of IMPI's 49th microwave power symposium, San Diego, USA, 16–18 June 2015

Chapter 2
Solid-State RF Power Generators

Roger Williams

Abstract This chapter is intended to provide an overview of the benefits that solid-state RF generators can provide to industrial, scientific, and medical (ISM) 'RF energy' applications, and the technologies, architectures, and design philosophies used in such generators. Our intention is to look at this from the end-user's viewpoint and to provide information that will be most helpful to

- researchers in a broad range of scientific and medical disciplines who want to learn how to use solid-state RF generation in their applications;
- designers with ideas for products based on technology from these disciplines who want to learn about the feasibility of a solid-state generator and the tradeoffs they will need to consider;
- ISM equipment manufacturers who are looking for an overview of solid-state generator technology to decide if and how they should get involved with it.

Solid-state RF power has been used for decades in 1–500 MHz ISM applications. However, power generation in the important 915 MHz, 2.45 GHz, and 5.8 GHz ISM bands has until recently been dominated by magnetrons, so those frequencies are the focus of this chapter.

2.1 Introduction

2.1.1 The Magnetron

It may seem strange to begin a chapter on solid-state RF generators with an introduction to the magnetron. However, the resonant cavity magnetron has dominated the fields of microwave heating and plasma generation for half a century, and it is valuable for users and designers of solid-state generators to understand the characteristics of the magnetron in order to appreciate what makes the solid-state source unique, and how to determine when it is a better solution.

R. Williams (✉)
3D RF Energy Corp,3 Davol Square, Providence, RI 02903, USA
e-mail: roger.williams@3drfe.com

S. Horikoshi and N. Serpone (eds.), *RF Power Semiconductor Generator Application in Heating and Energy Utilization*, https://doi.org/10.1007/978-981-15-3548-2_2

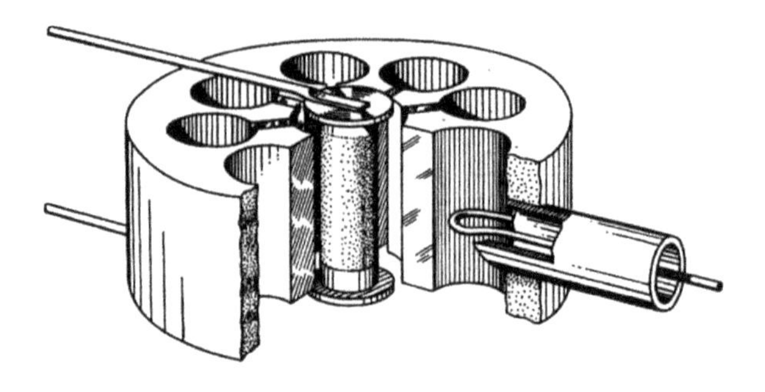

Fig. 2.1 Cutaway view of a 2.45 GHz cavity magnetron, showing the anode, cathode, and output coupling loop

The cavity magnetron is a high-power oscillator that generates microwaves using the interaction of a stream of electrons with a magnetic field (Fig. 2.1). To oversimplify the operating principle, the magnetron is a diode with a cylindrical cathode inserted coaxially in a cylindrical anode which contains a set of resonant cavities forming a slow wave structure [1]. Permanent magnets or electromagnets produce a static magnetic field along the axis of the cathode, while a high-voltage (4–50 kV) potential on the anode produces an electric field perpendicular to the magnetic field.

Electrons are ejected from the heated cathode by thermionic emission and are accelerated by the electrical field. They are deflected by the axial magnetic field into cycloidal paths past the cavities, pumping their natural resonant frequency and producing oscillating magnetic and electrical fields within the cavities. This oscillation in turn creates alternating charges following the oscillation on the internal surfaces of the anode segments between cavities. As electrons sweep towards a negatively charged point on the anode, they push that charge around the cavity, transferring their energy to the RF oscillating field. Power is extracted from this self-sustaining oscillation from one cavity, generally by a coupling loop.

This technology has evolved and matured into today's mass-produced consumer oven and industrial magnetrons.

So for the purposes of microwave power generation for ISM applications, a relevant question is what are the important characteristics of the industrial magnetron?

- Frequency stability is limited by loaded Q (typically 50–100) to $\pm$10 MHz—but it is sometimes feasible to injection-lock it to a lower-power solid-state source [2].
- It is fundamentally a constant power device, not a linear amplifier—but average power can be controlled over a wide range with a switched-mode power supply.
- Power stability $<1\%$
- Pulsed operation to $<25\,\mu$s, albeit at the cost of power supply complexity.
- Load VSWR (voltage standing wave ratio) $\leq$ 4:1
- Limited lifetime (typically 10,000 h)—but field replaceable.
- Frequency is controlled by geometry, so the minimum size is limited by frequency.
- The smallest practical size at 2.45 GHz is about 1 kW (9 dm^3 magnetron head)

To this, we can add some characteristics of a typical 75 kW, 915 MHz generator[1]:

[1]Muegge MH075KS-510CF head, CWM-75L magnetron, MX075KE-510ML power supply.

- Efficiency 78%, typical
- Cost 1–1.5 $/W
- Size of magnetron head $0.3\,m^3$, power supply $1.3\,m^3$
- Power density $47\,kW/m^3$.

It seems that the industrial magnetron is well suited as a compact, efficient power source for applications that need continuous or slowly pulsed power above 1 kW, can operate in the standard 915 MHz, 2.45 GHz, or 5.8 GHz ISM band, are uncritical about frequency stability, and are unconcerned about periodic magnetron replacement.

It is worth noting that the headwinds that have faced RF power semiconductor manufacturers in their engagement with RF energy are largely due to the fact that their motivation to find new high-volume markets for transistors has led them to focus on applications which are largely insensitive to the deficiencies of magnetrons, notably consumer cooking. Nevertheless, as we shall see, the benefits they have ascribed to solid-state generators are (mostly) valid and can be valuably and immediately applied to a wide range of ISM applications.

2.1.2 Benefits of the Solid-State Generator

The basic solid-state building block for power amplification and generation at RF and microwave frequencies is the RF power transistor. But before we look at details of its operation and use, let us take a high-level look at the potential benefits of solid-state RF generators for ISM applications.

Precise control of power, frequency, and phase Since solid-state generators are usually designed as a power amplifier (PA) driven by a low-power frequency synthesiser referenced to a crystal oscillator or external frequency reference; power level, frequency, phase, and modulation can be inexpensively and precisely controlled within that low-power exciter—often within the same RF integrated circuit.

Modulatibility In ISM applications, this usually means pulse modulation and amplitude control. But although it will be a rare ISM application that can benefit from the wideband, complex modulation that is ubiquitous in modern telecommunications, all solid-state generators benefit from the PA knowledge and thousands of small-signal RF components that have been developed for these applications.

Coherence When (as is usual) the signal source is a continuously running frequency synthesiser locked to a reference clock provided by a crystal oscillator or external clock, the generator output (a) is always coherent with the continuous synthesiser output and (b) can be designed to be phase coherent with the reference frequency.[2]

---

[2]Phase coherency is the attribute of two or more waves, or parts of a wave, where their relative phase is constant during the resolving time of the observer. It can also be interpreted for signals of different frequencies, where signals are at a specified phase relationship every N cycles. A simple example of this is when a PLL synthesiser's output frequency is an integer multiple of the reference frequency, e.g. a 2.450 GHz output is phase coherent with the 1 MHz reference.

A generator system can be composed of a number of lower-power self-contained generator modules, all locked to a common reference clock distributed from a central point or from one generator module to another. Since all of the module outputs are phase coherent to a common clock, if they are set to generate the same frequency, they will be coherent with each other as well. With due attention to phase alignment, their outputs can be constructively combined for higher power.

However, this capability may be more important in multichannel applications where the phase relationship between all channels is controlled to produce a particular desired effect. For example, in an application cavity containing multiple antennas, they can be driven coherently with the appropriate phase delay between each to produce beamforming (a uniform wavefront travelling in a specified direction) or adjusted to provide a more homogeneous power distribution in the cavity [3].

Coherence between the (gated and variable power) generator output and the continuous synthesiser output allows the generator to perform measurements such as real-time impedance of the application load; and phase coherence with a reference clock allows measurements to make use of powerful coherent sampling techniques. Coherent measurements are described in more detail in Sect. 2.4.5.

Long-term reliability Modern RF power transistors operated under normal conditions have a lifetime of more than 10^6 h, so the operational lifetime of a solid-state generator with properly designed PAs is not determined by the lifetime of its semiconductors. In practice, generator failures are most often the result of (a) inadequate protection (e.g. isolators, mismatch-insensitive combiners, arc detection, and control) against reflected power; (b) cooling system failure; (c) failures in electromechanical components (e.g. switches and connectors).

Stability at all power levels In addition to the frequency, phase, and power stability provided by the low-power exciter, we need to consider the stability of the PA itself. Because the gain of the transistor is higher at lower frequencies, the instability of RF transistors at low out-of-band frequencies was the source of much frustration among PA designers in the past.

Fortunately, the demands of the base station industry over the past 15 years have been such that a great deal of attention has been paid to the subject by transistor manufacturers. As a result, (a) potential instability is now addressed during the design of the die and internal matching networks, and (b) the electrical models provided by the manufacturers accurately model stability, providing a straightforward process for designing unconditionally stable amplifiers.

Compact, modular, scalable Every solid-state generator is a system, and most of them achieve high powers through a modular approach, combining the power from multiple identical power amplifiers or generators. It is worth remembering that—when automated assembly allows us to manufacture 1,000 identical PCB-based PAs as easily as one—'Design once, deploy everywhere'[3] is an effective strategy for constructing high-power generators.

[3]The network industry's somewhat tongue-in-cheek paraphrase of the 'write once, run everywhere' slogan created by Sun Microsystems to illustrate the cross-platform benefits of the Java language.

Rapid response Because of all of the operating parameters of a solid-state generator—frequency, phase, level, gating, transistor bias, etc.—are controlled by high-speed low-power circuits, changing the operating state in response to a sensed internal or external condition can be done as quickly as desired—in nanoseconds, if needed. This can be done for protection—reducing PA power in response to an externally detected arc, for instance—or to adjust generator frequency or power in response to changing application conditions.

Wide range of operating frequencies The RF power transistor is intrinsically a broadband device, although for practical reasons packaged devices are likely to use internal frequency-specific matching networks above 1 GHz. Gain and available power decrease with frequency, so while single transistors can deliver 2 kW at 100 MHz, the largest devices currently available for operation in the 5.8 GHz ISM band are capable of only about 200 W.

Powered by low voltage The 28–50 V operating voltage of RF power transistors allows the use of power supplies manufactured in high volumes for telecommunication and networking, which are cheaper and more compact than the rather specialised high-voltage switched-mode supplies designed for industrial magnetrons.

Cost parity with industrial magnetron systems The price of RF power transistors used in the 915 MHz and 2.45 GHz ISM bands has fallen dramatically over the past few years, so the total cost of a solid-state generator is unlikely to be dominated by transistor cost. Still, a solid-state generator contains many more components than a magnetron generator, and at this early stage in this market, a solid-state generator is still likely to be twice as expensive as an industrial magnetron generator of equivalent power. However, when we consider the entire application, we can frequently eliminate or simplify system components that one or more unique characteristic of solid-state generators (e.g. frequency agility) renders unnecessary.

2.1.3 The Need for a Systems Approach

Figure 2.2 shows the top-level block diagram of a commercial 75 kW, 915 MHz solid-state generator. Each of the 2.5 kW PA modules is an independent hot-swappable subsystem with its own monitoring and control. In turn, each of those modules contains four 650 W PA modules (constructed as PCB pallets) and associated drivers, splitter, combiner, and isolator.

Despite its apparent complexity, we believe that the development of a solid-state RF generator can be quick and successful when a system's approach is taken. By this, we mean a holistic approach that spans the end application, the range of technologies and design disciplines used, and the life span—from concept definition to design to manufacturing to operation and support—of the generator.

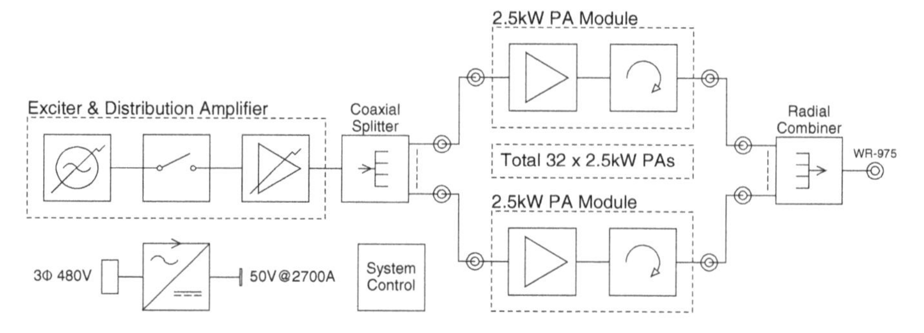

Fig. 2.2 Simplified block diagram of solid-state 75 kW, 915 MHz RF generator

The sad fact is that 'systems engineering' in many institutions has become siloed instead of holistic. We are reminded of the parable[4] [4] about the blind men feeling different parts of an elephant and deducing different animals. The development of multi-disciplinary products will be most successful when each participant understands each other's application domains and can see the big picture perspective.

2.2 RF Power Semiconductors

The first RF transistor appeared at the end of the 1960 s, a silicon BJT (bipolar junction transistor) that could deliver 50 W for HF transmitter applications (Fig. 2.3). The power and frequency limits of RF power BJTs continuously improved for the next 30 years, even as competing technologies arrived on the scene.

The first of these was the V-groove metal oxide semiconductor (VMOS) field-effect transistor introduced in 1979 [5], capable of producing 300 W at 200 MHz. This technology was improved by changing the device to a vertical double-diffused MOS (VDMOS) structure by the end of the 1980s. VDMOS has a number of advantages over BJT—simpler biasing, higher gain, better thermal stability, and superior load mismatch tolerance—but its larger RF parasitics (feedback capacitance and source inductance) practically limited its use to below 500 MHz. The BJT remained the only technology for UHF and microwave communication PAs.

The reign of the RF power BJT ended in the late 1990s when laterally double-diffused MOS (LDMOS) transistor technology was introduced as a replacement for BJT in base station applications, which it completely dominated within a few years. VDMOS transistors continue to be manufactured today, but they have been replaced by LDMOS in all but niche applications, such as plasma generators $\leq$ 30 MHz.

[4]The original fable originated in China during the Han dynasty [4]. Although India and Africa have similar stories, the philosophical note is admirably Confucian: When a person is opinionated or blind to his limitations because of insufficient knowledge or smug mentality, he is as blind as if he had no eyesight.

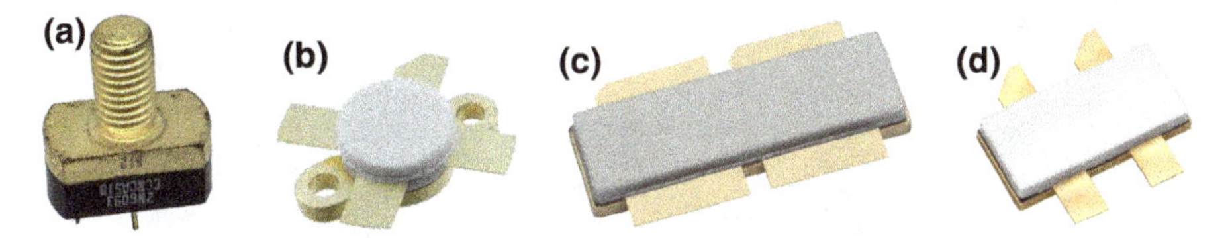

Fig. 2.3 50 years of RF power transistors: **a** 50 W, 30 MHz BJT ca. 1969; **b** 20 W, 160 MHz VMOS ca. 1976; **c** 750 W, 915 MHz LDMOS ca. 2017; **d** 330 W, 2.45 GHz GaN ca. 2019

Around 2005 [6], the first commercial RF power devices manufactured with GaN technology were introduced to the market. Despite significantly higher manufacturing costs, the GaN high electron mobility transistor (HEMT) has one particular characteristic that immediately endeared it to the military communication and radar markets: a power density 4 times that of LDMOS, which results in parasitic capacitances 1/4 that of LDMOS. This allows high-power operation at higher frequencies (>10 GHz) and over very broadbands (e.g. 0.1–3 GHz at 200 W), enables switched-mode classes of operation >1 GHz, and simplifies the design of narrowband amplifiers >3 GHz.

2.2.1 *LDMOS*

Today, LDMOS is the leading technology for a wide range of RF power applications up to 4 GHz, including base stations, broadcast, and ISM. This is even true at 2.45 GHz, despite the recent interest by some GaN manufacturers in RF energy markets. The reason for this has been the remarkable increase in LDMOS performance and reduction in cost over the past 15 years (Fig. 2.4).

This has been driven by the high-volume cellular base station market, where fierce competition has fuelled continuous improvement of the technology and reduction of manufacturing cost. Of all RF power technologies, LDMOS has the unique advantage of being fabricated in a standard CMOS fab, and the ability to exploit the wafer manufacturing and lithography technology that has been developed and refined by the $500B CMOS industry.

In addition to the ability to take advantage of the low-cost manufacturing technologies developed for CMOS, the underlying reason for such rapid performance improvement in LDMOS (and hence the rapid transition from BJT to LDMOS) is that LDMOS is a *surface dominated* device, allowing easy manipulation of its FET characteristics during the manufacturing process without modifying the underlying epi. This is particularly important for optimisation of device characteristics (parasitic capacitance and inductance, drain extension, etc.) critical to RF performance.

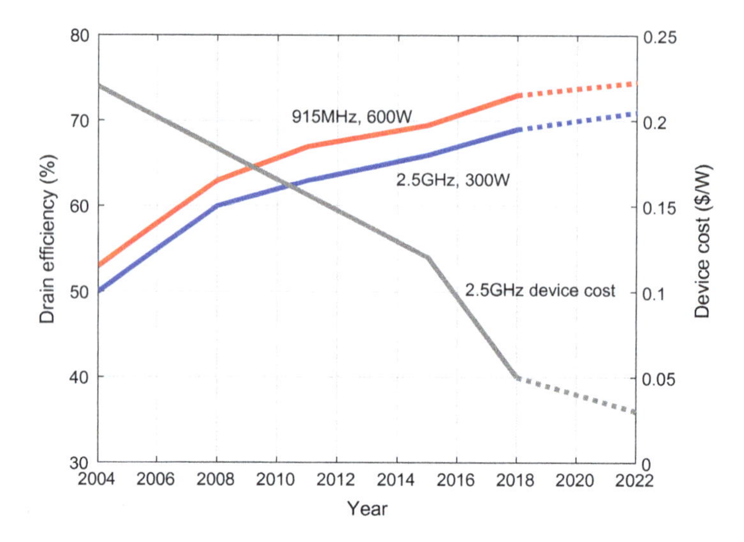

Fig. 2.4 Trends in LDMOS device efficiency and cost

2.2.1.1 LDMOS Theory of Operation

Like other MOSFETs, the LDMOS transistor is a field-effect transistor, where the voltage on the (oxide-insulated) gate controls the current flowing from the drain to the source; and like them, it was made possible by the evolution of CMOS technology. Unlike other power MOSFETs, it is a lateral structure that achieves high current handling through a massively parallel structure.

Although processed in a CMOS fab, it is differentiated from a typical CMOS FET with features that allow it to operate at higher voltage (which it needs to generate high RF power), primarily a drain extension region and laterally diffused channel doping to control gain and threshold. Other features provided for RF performance and reliability include a source sinker to the substrate to reduce R_s and L_s, backside metallisation to improve heat transfer, and grounded tungsten shields which help to control the E-fields near the drain edge of the gate, reducing C_{gd} and improving hot carrier reliability properties [7].

Figure 2.5 shows a schematic cross section of an LDMOS transistor and photograph of a packaged device. Its n+ source region is connected to the backside via a metal sinker, a p+ sinker, and a highly conducting p+ substrate. Electrons flow from the source to drain if the gate is positively biased, inverting the laterally diffused p-well channel. Many fingers are placed in parallel to form a power die, resulting in a total finger length of 10–1000 mm.

Fundamentally, the LDMOS transistor is a voltage-controlled current source, as illustrated in Fig. 2.6. DC analysis (Fig. 2.7) provides the drain–source breakdown voltage BV_{dss}, maximum current I_{dsx}, on-resistance R_{on}, knee voltage V_k, threshold voltage V_{th}, and transconductance $g_m = dI_{ds}/dV_{gs}$. The usual small-signal figures of

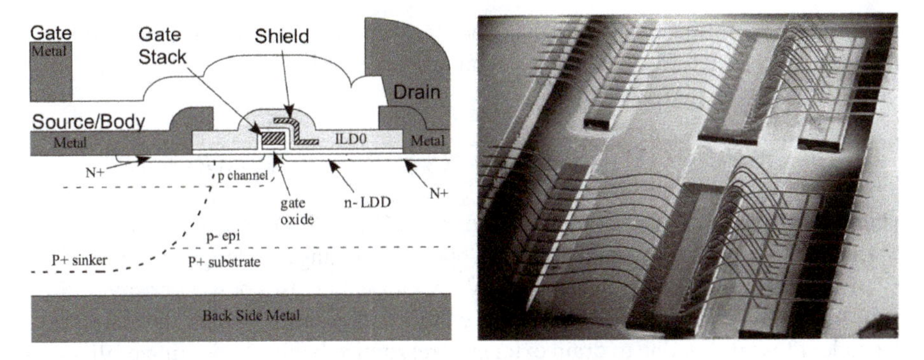

Fig. 2.5 LDMOS cross section and detail of the packaged device. The photograph shows (left to right): gate lead, input low-pass match with bond wires and MOS cap, LDMOS die, output match with bond wire shunt inductor and MOS cap DC block, drain lead

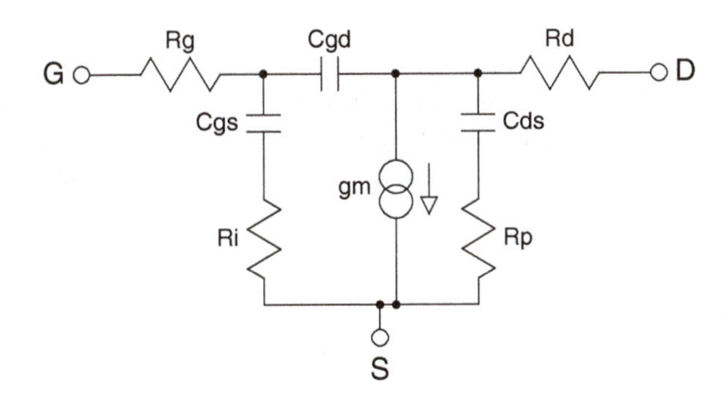

Fig. 2.6 LDMOS small-signal equivalent circuit

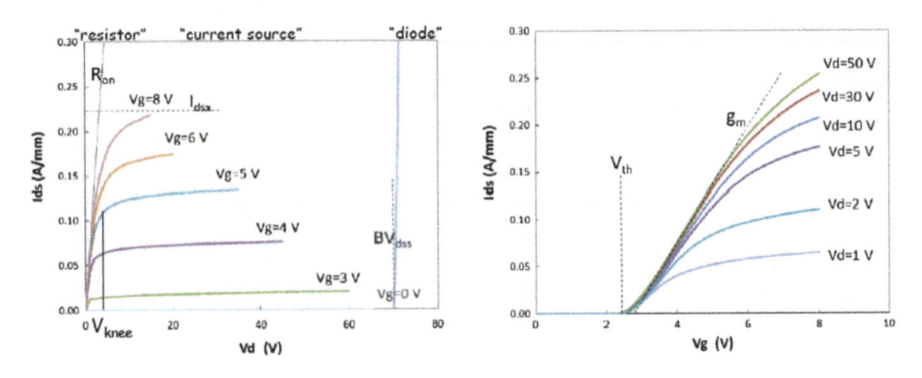

Fig. 2.7 LDMOS I-V curves

merit for RF performance, transit frequency f_T, and maximum oscillation frequency f_{max} can be obtained from the small-signal model as [8]

$$f_T \approx \frac{g_m}{2\pi(C_{gs} + C_{gd})} \quad \text{and} \quad f_{max} \approx \frac{f_T}{2\sqrt{g_{ds}R_g + 2\pi f_T C_{gd}(R_g + \alpha R_d)}}$$

where g_{ds} is the channel conductance and $\alpha = (C_{ds} + C_{gd})/(C_{gs} + C_{gd})$.

LDMOS power density increases with operating voltage, so lengthening the drain extension to increase BV_{dss} and allow 50 V operation increases power density by 50% from that at 30 V. However, as the drain extension becomes longer, the frequency-dependent loss in C_{ds} due to drain extension resistance begins to dominate [9], practically limiting efficient 50 V operation to 1 GHz.

2.2.2 *GaN*

In the early 1990s, GaN was identified as the best material for the next generation of high-frequency, high-power transistors, due to its high bandgap, electron mobility, and saturated electron velocity. However, the need for fundamental material development limited GaN RF transistor development. This restraint was finally eliminated by the intense material development that resulted from the race to first production of blue LEDs—the availability of device-grade AlGaN/GaN heterostructures finally allowed the demonstration of high-power GaN RF transistors. Although high-power LEDs commonly use sapphire as a bulk substrate, GaN transistors usually use SiC because of its higher thermal conductivity.

About five years ago, it became apparent that many of the emerging requirements for 5G cellular base stations—new frequency bands from 3.5–7.1 GHz and new wideband requirements below 3 GHz—can only be achieved with GaN transistors. As a result, the GaN industry has entered a very competitive race to improve performance and develop high-volume manufacturing technologies that mirror the remarkable growth of the LDMOS industry (driven by 3G and 4G cellular demands) 15 years earlier.

For example, in the past three years the industry has seen an explosion in the number of GaN device manufacturers and GaN-on-SiC epi wafer suppliers, the introduction of 150 mm wafers, and the development of low-cost, high-volume packages (e.g. 300 W plastic wire-bond-less packages for less than 1/3 the cost of standard air cavity ceramic).

2.2.2.1 GaN HEMT Theory of Operation

The HEMT is also a FET, i.e. a voltage-controlled current source. However, its operating principle is somewhat different from the MOSFET. Instead of using a

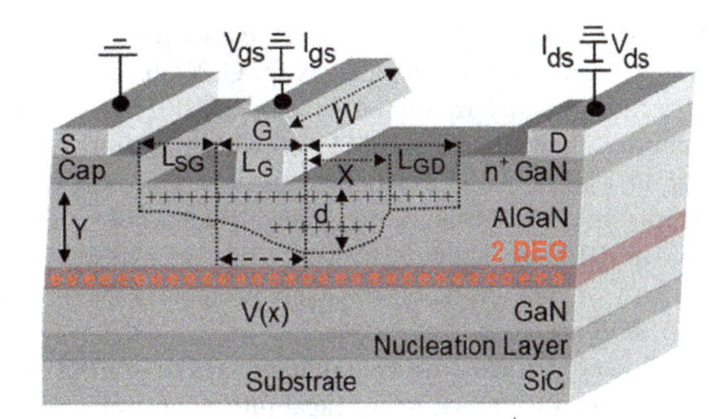

Fig. 2.8 Cross section and operation of an AlGaN/GaN/SiC HEMT biased under a common source configuration

doped region, the HEMT uses a heterojunction, a junction between two materials with different bandgaps: AlGaN and GaN in this case [10].

A quantum well is formed in a narrow gap at the interface between the two materials, and electrons from the wider bandgap material (AlGaN) fall into this potential well and are confined there. Because of quantum mechanical confinement in such a narrow dimension, they form a high-density electron gas in two dimensions: although they can move freely within the plane of the heterointerface, perpendicular movement is restricted. This so-called two-dimensional electron gas (2DEG) is the conductor between drain and source (Fig. 2.8).

Because of spontaneous polarisation (due to lack of symmetry) and piezoelectric polarisation (due to the different lattice constants of GaN and AlGaN), a sheet of positive uncompensated charge is formed, causing the 2DEG to be formed even without doping.[5] Thus, the usual GaN HEMT is a *depletion* device: it is normally *on* and will turn off only when the gate is negatively biased.

Because current is carried by the 2DEG, the drain and source must make ohmic contact with the AlGaN/GaN heterostructure. The gate is formed by a Schottky contact, i.e. a metal–semiconductor contact with a large barrier height and low doping concentration. This forms a barrier for electrons and holes (i.e. an insulator), but also leads to gate leakage and RF rectification currents (Sect. 2.3.5.2) not present in MOSFETs.

2.2.2.2 GaN-on-Si

SiC does not melt, but instead gradually sublimes (200–500 μm/h) at a process temperature of 1800–2400°. Thus, it isn't possible to form large single-crystal ingots by pulling a seed crystal from a melt—at up to 200 mm/h—as in the Czochralski process used to produce 200 and 300 mm diameter silicon ingots.

[5] By doping the barrier with acceptors such as Mg, this built-in charge can be compensated to restore enhancement mode. However, this process introduces effects (such as trapping) which degrade high-frequency power and efficiency, so the p-GaN HEMT is currently used only for high-voltage switching devices.

As a result, many semiconductor researchers concluded that there would never be a clear commercialisation pathway for GaN-on-SiC and began to develop GaN grown directly on standard silicon wafers. In 2001, researchers at Nitronex Corporation reported their first RF HEMTs grown on 100 mm Si wafers, and in 2006 shipped the first 50 W, 3.3–3.8 GHz devices [11]. Unfortunately, despite beginning volume shipments in 2009, they struggled with technology, manufacturing, and capital issues and in 2014 were acquired by MACOM. Since then, MACOM has continued to invest heavily in GaN-on-Si, aggressively promoting the message of 'GaN performance at LDMOS cost'.

However, since 2014 the commercial GaN-on-SiC industry has also evolved, the improvements in processing capability, supply chain robustness, and manufacturing processes have significantly reduced the cost of packaged GaN-on-SiC transistors. At the moment, it is unclear whether GaN-on-Si will be a significant transistor technology in the future.

2.2.3 Reliability and Thermal Behaviour

Semiconductor lifetime is specified as median time to failure (MTF), the time that 50% of the population has failed.[6] For LDMOS transistors, this is a function of junction temperature (T_j) of the device, assuming electromigration as the wear-out failure mechanism; for GaN HEMTs, it is a function of surface temperature (T_s), assuming Schottky and ohmic degradation as the wear-out mechanism.

These wear-out mechanisms follow Black's equation [13], so MTF $\propto e^{1/T}$. For example, a 750 W, 915 MHz LDMOS transistor (BLF0910H9LS750P) delivering the full rated 750 W of power dissipates 284 W, which for a case (mounting flange) temperature of 85° results in a junction temperature of 128° and a calculated MTF of 6×10^6 h (691 years)—but this decreases by a factor of 2 with every 10° increase in temperature.

Because the efficiency of RF power amplifiers at 915 MHz and 2.45 GHz is only 65–75%, they dissipate 1/4 to 1/3 of the DC power as heat. RF performance (e.g. power, gain, efficiency) degrades as junction temperature increases. Since lifetime is a strong function of temperature too, it is critical to design the die and package for a low thermal resistance from the LDMOS junction (or GaN HEMT channel) and the mounting surface of the packaged transistor.

This is easier with LDMOS than GaN: the higher power density of GaN inevitably results in higher die surface temperatures because of the much smaller die area,[7]

---

[6]MTF is statistically distinct from mean time to failure (MTTF). In modern LDMOS, high-temperature electromigration is the dominating deterioration failure mechanism. Since this has a lognormal distribution [12], MTTF $\propto e^{\mu+\frac{1}{2}\sigma^2}$ is slightly higher than MTF $\propto e^{\mu}$.

[7]Yes, the thermal conductivity of SiC is 3:1 that of Si at room temperature, but drops to 2:1 at 150°. Moreover, modern LDMOS typically uses 50 μm Si substrates compared to GaN's current 100 μm SiC substrates.

resulting in typical 2:1 differences in $R_{\theta JC}$ between comparable GaN and LDMOS transistors. Fortunately, GaN's higher maximum temperature (300–350° HEMT channel versus 225–250° for LDMOS junction) generally compensates for this.

Both LDMOS and GaN die are attached to a copper alloy flange in the package by eutectic solder or sintered epoxies. In turn, this flange is generally soldered to a copper coin or baseplate in the power amplifier PCB structure.

2.2.4 *Ruggedness*

The most common ruggedness failure mechanism for LDMOS transistors is a catastrophic failure resulting from the snapback of a parasitic BJT formed by the drain (collector), channel (base), and source (emitter) of the LDMOS (Fig. 2.9) [14]. If the BJT is turned on, it creates a very low resistance path between source and drain in a very small discrete location (i.e. one finger) of the LDMOS die. The resulting current flow and thermal dissipation that occurs vaporises the localised silicon and associated metal interconnects, which can then cascade into a series of other failures resulting in total destruction of the die.

Snapback is initiated by a small current from collector to base. When the LDMOS is not conducting (i.e. the positive cycle of the V_d waveform), the drain–source diode normally clamps the voltage across the LDMOS (and BJT) to the substrate. But if a large enough sink current is applied to the drain, the drain–source voltage exceeds the diode breakdown voltage and the BJT can be triggered. This threshold is typically about 25% above BV_{dss}, e.g. about 80 V in 30 V LDMOS with $BV_{dss} = 65$ V. Once the BJT turns on, the collector voltage decreases to the snapback holding voltage.

Large sink currents can be caused by a mismatch, improper harmonic termination, or operation in saturation. The resulting ionisation breakdown can provide a sufficient source of charge carriers injected into the base to turn on the BJT. Because charge can be injected through C_{db}, it can also be triggered by a high dV_{ds}/dt, once V_{ds}

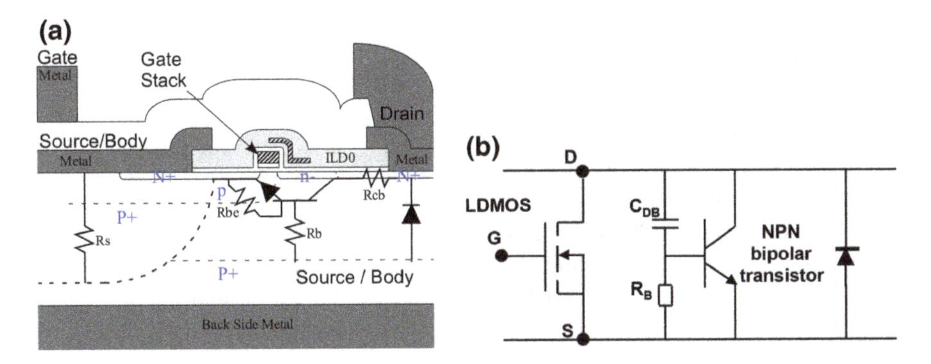

Fig. 2.9 LDMOS parasitic BJT: **a** cross section with BJT overlayed; **b** equivalent circuit

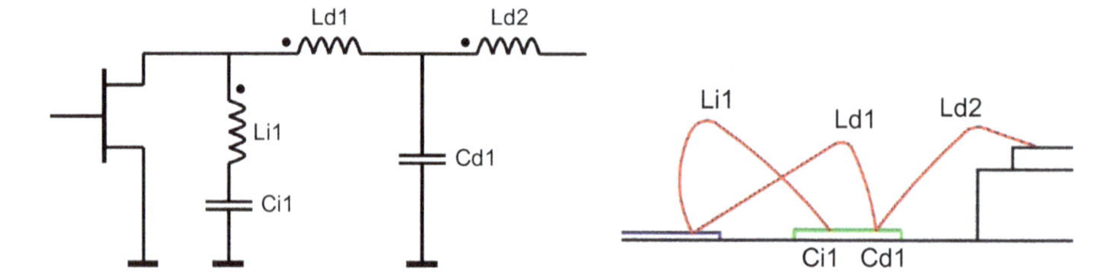

Fig. 2.10 Internal output matching network for 250 W, 2.45 GHz LDMOS transistor. This network uses an internal shunt inductor (Li1) with DC block (Ci1) to compensate for C_{ds}, followed by a low-pass network (Ld1, Cd1, Ld2) to transform the 1.4 Ω R_L to about 3 Ω at the drain lead

is enough higher than BV_{dss}. Although R_b and C_{db} are optimised in the LDMOS process, it is still important to consider snapback triggering during the design of PAs that operate at saturation with pulsed RF signals (Sect. 2.3.6).

2.2.5 Internal Impedance Matching

As described in the appendix at the end of this chapter, the load impedance of an RF transistor is primarily a function of the supply voltage and the desired output power. For high-power devices, the input impedance is also low as a result of the large die needed, but is dominated by gate capacitances. To make it practical to use them in a PCB-based power amplifier, such high-frequency transistors are usually 'prematched' to higher impedance at the transistor leads with internal matching networks based on bond wire inductors and MOS shunt capacitors integrated into the transistor package. Figure 2.10 shows the circuit and physical layout of an output matching network for an LDMOS device.[8] The input is matched in the same way, usually with a single or double low-pass network.

2.2.6 Simulation Models

Although small-signal equivalent circuits are valuable for understanding the operation of the transistors, models useful for power amplifier design need to accurately model large-signal, nonlinear behaviour including dynamic self-heating effects. However, large-signal behaviour at high frequencies is too complex to practically model entirely as a physical model, so over the years various 'compact' models (these days usually written in Verilog-A) for use in circuit simulation have been

[8] Because the GaN transistors used for 2.45 GHz power applications are usually 50 V devices, R_L and C_{DS} are 'friendly' enough (e.g. 3.7 Ω and 25 pF for a 300 W device) that internal output matching networks are not required.

developed [15, 16] in which the behaviour of physical components is described by non-physical fitting functions. Alas, the details of these are closely guarded by the transistor manufacturers, who supply the models for their devices exclusively as encrypted libraries for Keysight ADS and National Instruments AWR commercial simulators.

2.3 RF Power Amplifier Design

There is not enough space in this chapter for any RF power amplifier design tutorial that would do justice to the subject—indeed, this could (and does) occupy entire books. Steve Cripps in particular does an excellent job [17] at tackling difficult topics and providing a thorough but easy-to-understand explanation. Nevertheless, understanding the underlying principles and tradeoffs of PA design is essential for the generator designer and—we hope—interesting to the end-user.

2.3.1 Key Performance Parameters

For amplifiers used for RF power generation, we are principally concerned with three parameters describing RF performance, shown in Fig. 2.11, as well as ruggedness and thermal behaviour. For each of these, read the datasheet to determine if the parameter is specified under CW (continuous) or pulsed operation. Since the RF parameters are a function of frequency, temperature, and supply voltage, it is important that the datasheet also provide curves showing those dependencies.

Power gain $G_P = P_{\text{out}}/P_{\text{in}}$ is usually specified at the PA's nominal power rating, but may also be specified at a given compression point, or even as peak gain. Gain is a strong function of temperature, typically $-0.02\,\text{dB}/^{\circ}$.

Efficiency is expressed as 'drain efficiency' $\eta_D = P_{\text{out}}/P_{dc}$ or 'power added efficiency' PAE $= (P_{\text{out}} - P_{\text{in}})/P_{dc}$.

Compression describes the characteristic of an amplifier to become increasingly nonlinear as output power increases above a certain level. Output power is specified at a specific 'compression point', the point at which the gain has decreased from its peak value by a given amount, e.g. $P_{L(1\,\text{dB})}$ is the output power at the point where gain has dropped by 1 dB. Although the PA is extremely nonlinear when in gain compression, the CW efficiency of a class B amplifier is highest at 1–2 dB compression.

Ruggedness describes the capability of a transistor or PA to survive a large output mismatch, expressed by the load VSWR (voltage standing wave ratio). Since the phase of the reflected power determines whether the transistor sees a voltage maximum or current maximum, it is important that the allowable mismatch is specified

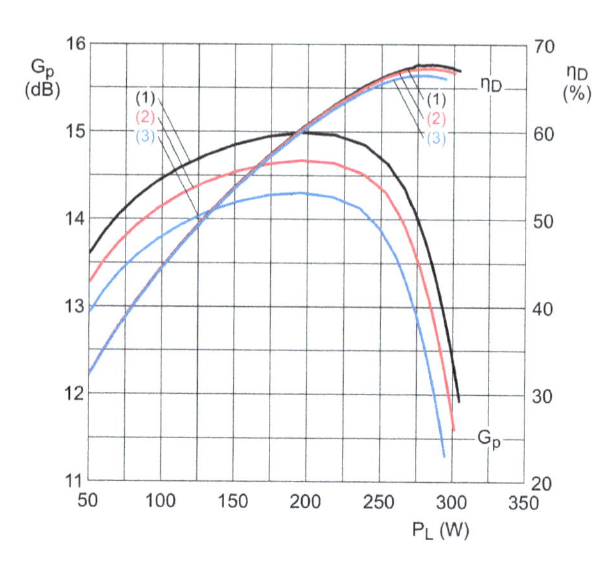

Fig. 2.11 Power amplifier drive-up curves describe power gain, efficiency, and gain compression of a 250 W, 2.45 GHz, 32 V LDMOS transistor (BLC2425M10LS250) in a specified test circuit. Gain and efficiency curves are plotted for case temperatures of (1) 25°, (2) 45°, and (3) 65° to show the temperature dependency of these parameters

across 'all phases'. Note that an integrated or modular PA is often not as rugged as the transistor it uses, due to lower voltage or current breakdown limits in other components (e.g. capacitors), the output connector, or even the PCB substrate.

Thermal resistance between transistor junction and device mounting plane. Thermal management is described in more detail in Sect. 2.4.6.

2.3.2 Power Amplifier Classes

Power amplifiers are differentiated from small-signal amplifiers by their circuit configurations and methods of operation. They are identified by their classes of operation, that is, classes A, B, C, D, E, F, etc. (Classes D, E, and F are often referred to as 'switched-mode' classes, since the output device(s) are operated as switches rather than as linear elements. Class D uses pulse width modulation to control the conduction angle of the output devices. Although popular at lower frequencies, high-power class D amplifiers are impractical above about 30 MHz.) Because power generator efficiency is such an important parameter, PA designers pay a great deal of attention to the 'high efficiency' classes, i.e. C, D, E, and F. Table 2.1 summarises the characteristics of the PA classes used in the 915 MHz and 2.45 GHz ISM bands.

Classes B and AB (in which a conduction angle slightly more than 180° is used to improve gain and linearity) have dominated high frequency solid-state RF PA designs because of their reasonable power, gain, efficiency, and linearity, and the possibility to adjust the output load and operating point to 'tune for' (selectively optimise) one or two of these parameters to best suit the application.

Table 2.1 Power amplifier classes of operation

Class	Features	P_{max}[a,b]	η_{max}[b]	Implementations
A	FET in active region at all times, best linearity V_D, I_D are both sinusoids Highest gain	0.25	50%	
B	FET active 180° (V_{GS} at threshold), linear I_D is half-sinusoid, V_D is sinusoid Most common RF power amplifier class	0.125	78.5%	700 W, 74% @ 915 MHz 500 W, 70% @ 2.45 GHz
C	FET active $\leq$ 180° (V_{GS} < threshold), nonlinear Lower gain, higher peak voltage than class B[c] V_D is sinusoid, $I_D \approx$ narrow pulse[d] Widely used in high-power vacuum tube PAs	0.10@3f$_0$ 0.08@5f$_0$	81%@3f$_0$ 90%@5f$_0$	200 W, 82% @ 433 MHz
E	FET operated as switch, nonlinear Even & odd harmonics in both V_D and I_D[d] Lower gain, higher peak voltage than class B	0.10@3f$_0$ 0.10@5f$_0$	81%@3f$_0$ 90%@5f$_0$	40 W, 82% @ 2.45 GHz
F	FET operated as switch, nonlinear Harmonic resonators at output to shape V_D, I_D $V_D \approx$ square wave, $I_D \approx$ half-sinusoid[d] Lower gain than class B Most promising switched-mode class for 2.45 GHz	0.14@3f$_0$ 0.15@5f$_0$	81%@3f$_0$ 90%@5f$_0$	100 W, 73% @ 2.45 GHz

[a] Figure of merit for the power-output capability [18]. $P_{max} = P_{out}/(V_{D(max)} I_{max})$
[b] P_{max}, η_{max} calculated for even and odd-order harmonic terminations up to the specified order.
[c] Transistor manufacturers have largely ignored class C operation above 1 GHz, in part because of the focus on linearity for base station applications, and in part because of its lower gain and P_{max}. However, new GaN devices designed for 2.45 GHz ISM applications are often operated class C to improve efficiency. According to its manufacturer, one such example (MRF24G300H) typically delivers P_L = 330 W with η = 73% and G_p = 15.2 dB at V_{dc} = 48 V [19].
[d] An 'inverse' class exists, in which V_D and I_D waveforms are swapped

Class F is currently seen as the most promising class for power generation above 1 GHz, as it is the only one which can in theory achieve both higher power capability and higher efficiency than class B. Like the other switched-mode classes, however, it places great demands on both the high switching speed and low parasitic capacitances of the output device, and as a result, only the GaN HEMT is suitable.

2.3.3 Power Amplifier Packaging

Solid-state power amplifiers in our bands of interest are conventionally implemented with two-sided or multilayer PCB structures. Because of its simplicity and low cost, microstrip networks are commonly used for impedance matching, although stripline is often used for coupled lines and other structures requiring shielding. Shunt capacitors are frequently used to reduce the length of microstrip lines used for impedance transformation by forming low-pass pi networks. (Below 900 MHz, discrete inductors and a variety of transformers implemented with wire and coaxial cable are common, but they are rarely used in the 915 MHz ISM band and higher.)

The source terminal of a high-frequency RF power transistor is connected to its mounting flange, so the PCB contains an embedded copper coin, a thick metal backing, or is directly mounted on a metal baseplate to provide both a low-inductance source connection to ground and the cooling path for the transistor.

After thermal management, an important electromechanical structure needed for every PA is EMI shielding. This is not necessarily for meeting regulatory radiated emission limits—although those need to be considered in the generator design—but to keep RF fields from interfering with other electronics and to prevent feedback. The latter is of particular concern, as a PA module might have as much as 60 dB of gain—and all that is needed to turn a power amplifier into a power oscillator is unity gain at the right phase. (This means that a multi-stage PA will need shielding *between* stages as well.) Shielding is sometimes integrated, e.g. as machined cavities in an enclosure. More often though, this is provided by sheet metal shields soldered to the PCB and bolted-down machined or die-cast shields.

For the same reason, it is critical to provide EMI *filtering* on all control and power lines into the PA, lest those turn into antennas outside the shield. Power amplifier demo boards from transistor manufacturers are valuable for evaluating RF performance and experimenting with tuning, but they do *not* include everything needed for a manufacturable PA.

2.3.4 Impedance Matching

As described in Sect. 2.2.5, RF power transistors used above 1 GHz usually contain internal prematch networks to ease impedance matching on the PCB and save board space. However, it is still necessary to design matching networks between these ‘pre-matched’ impedance at the transistor leads and the PA input and output impedance (typically 50 Ω).

f (MHz)	Z_{source} (Ω)	Z_{load} (Ω)
2400	2.55 – j2.96	2.41 – j3.12
2450	2.55 – j2.72	2.13 – j2.98
2500	2.56 – j2.49	1.88 – j2.80

Fig. 2.12 Input and output impedance data for a 300 W, 2.45 GHz GaN transistor. These describe the impedance to be presented to the transistor

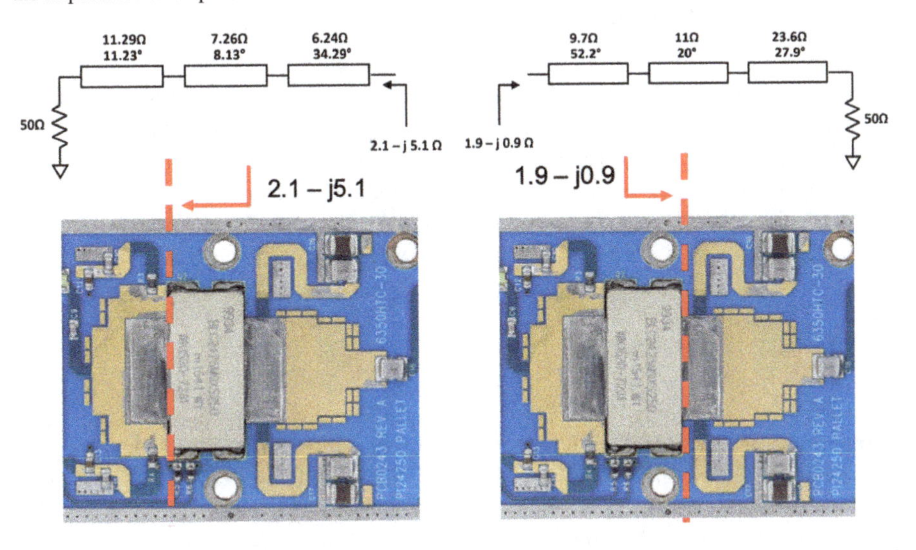

Fig. 2.13 Input and output matching networks for a 250 W, 2.45 GHz transistor, using a series of microstrip lines as stepped transformers. Ideally the drain power feed lines are $\lambda/4$ long so that they short-circuit even harmonics and appear open at the fundamental, but can be adjusted to fine-tune transistor reactance compensation

The impedance that the PA designer needs to present to the transistor is specified by the manufacturer in several ways: first, as a list of source and load impedance (Fig. 2.12) determined by load pull[9] for a specific application; second, as implicitly defined in the (ADS or AWR) device model, allowing the designer to optimise the match for the combination of parameters she desires; and finally, as expressed in hardware in the form of an evaluation board.

Figure 2.13 shows typical microstrip networks, these matching the input and output of a 250 W, 2.45 GHz LDMOS transistor to the 50 Ω PA terminals. Both input and output networks use three short microstrip lines of increasing impedance as simple stepped transformers to perform the ≈ 1:20 impedance transformation. DC power is fed to the drain through the two microstrip stubs: for class B operation,

[9] Load pull is an automated process of discretely varying the load impedance presented to a device under test while recording RF performance and optimum input impedance at each point. The measured data is typically used to generate contour maps of G_p, P_{max}, and η versus Z_L.

these are typically chosen to be $\lambda/4$ long so that they appear open at the operating frequency, but short-circuit even harmonics. These networks were designed with a Smith chart and published impedance data. The PA so designed met or exceeded the transistor manufacturer's 'typical' RF performance with only minor repositioning of the decoupling capacitors at the ends of the drain feed stubs.

2.3.5 *Bias and Control*

Nearly every PA (the classic exception is class C) needs stable low voltage sources to 'bias' the gates of the RF transistors to set 'quiescent' (no RF drive) drain currents. In class B/AB amplifiers used for RF power generation, we usually set the bias voltage V_{gs} slightly above gate threshold $V_{gs(\mathrm{th})}$ to provide a small quiescent drain current I_{dq}, typically 0.25–1% of $I_{dc(\mathrm{max})}$. For example, for a 250 W, 2.45 GHz LDMOS transistor that draws 12 A at full power we might set I_{dq} to 50–100 mA.

2.3.5.1 LDMOS Biasing

For LDMOS transistors, $V_{gs(\mathrm{th})}$ is typically 1.5–3 V. However, this threshold has a temperature coefficient of about $-2\,\mathrm{mV}/^\circ$ (referenced to the junction temperature), so temperature compensation must be provided for the gate bias voltage. It is feasible to monitor temperature with a small-signal diode or transistor and use an active bias generator [20] to precisely control the bias voltage. Some systems rely on the system controller to monitor PA temperature and continually adjust the bias voltages.

However, for power generation applications, we prefer the combination of a non-volatile DAC to set the bias voltage, with a thermistor mounted on the PA PCB near the power transistor to provide temperature compensation. This is shown in Fig. 2.14. (It is generally a good idea to put the DAC on a separate control board, not the PA PCB, because that PCB gets hotter than the small-signal PCBs.) This has the advantages of simplicity, always having correct PA bias levels regardless of the state of the system controller, and insensitivity to RF fields within the PA.

The slope (V/T) of the bias voltage needed to compensate for the $V_{gs(\mathrm{th})}$ temperature coefficient depends upon how closely the thermistor temperature tracks the transistor junction temperature, which is application and layout dependent. This is most easily determined empirically (e.g. by changing the baseplate temperature of an operating PA and measuring the resistance of R_{T} and the required value of V_{gs} for the desired I_{dq} at several different temperatures). The values of resistors R_1, R_2, and R_3 will then have to be calculated for the required V_{gs} slope. But once set up for a particular PA PCB layout, this circuit provides a sufficiently linear voltage slope as a function of junction temperature across a 100° ambient temperature range.

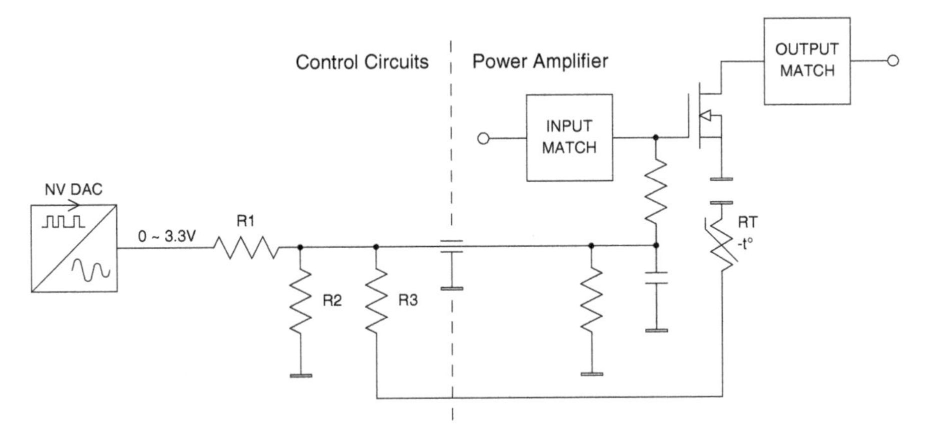

Fig. 2.14 Thermistor-compensated LDMOS bias network. NTC thermistor R_T is mounted near the transistor, and as part of the network with resistors R_1, R_2, and R_3 generates a voltage which is close to a linear function of temperature

2.3.5.2 GaN Biasing

GaN HEMTs present a more complicated situation. Because the HEMT is a *depletion* mode device, it will draw destructive drain currents from the DC supply unless the gate is biased to a sufficiently negative voltage. There are two issues to consider:

- Never apply drain voltage when the gate is at 0 V, as the device will draw excessive drain current. Thus, the bias source must include sequenced drain voltage switching.
- For a given V_{gs}, GaN HEMTs are prone to be potentially unstable at lower V_{ds}. Therefore, decrease the gate voltage to below the pinch-off voltage V_p while the drain voltage is being turned on and off.

Figure 2.15 illustrates recommended power-up and power-down sequences for the drain and gate voltages of a GaN HEMT [21].

In addition, because the GaN HEMT gate terminal is a Schottky diode, bias generators must provide significant amounts of both positive and negative gate current:

- GaN HEMTs have higher gate leakage currents than comparable LDMOS devices. The negative gate current can be as high as 500 μA/mm of gate periphery at elevated junction temperatures, e.g. −5 mA for a 100 W device operating at 200° surface temperature.
- When the device is driven into saturation, rectified positive gate current flows into the gate diode. At heavy RF compression, this gate current can exceed 1 mA/mm of gate periphery, e.g. 30 mA for a 100 W device.

The gate threshold voltage $V_{gs(\mathrm{th})}$ shifts positively with temperature [22], with a typical slope of 1–3 mV/° (depending upon device geometry and Schottky barrier height), so temperature compensation of gate voltage is normally also required for GaN HEMTs.

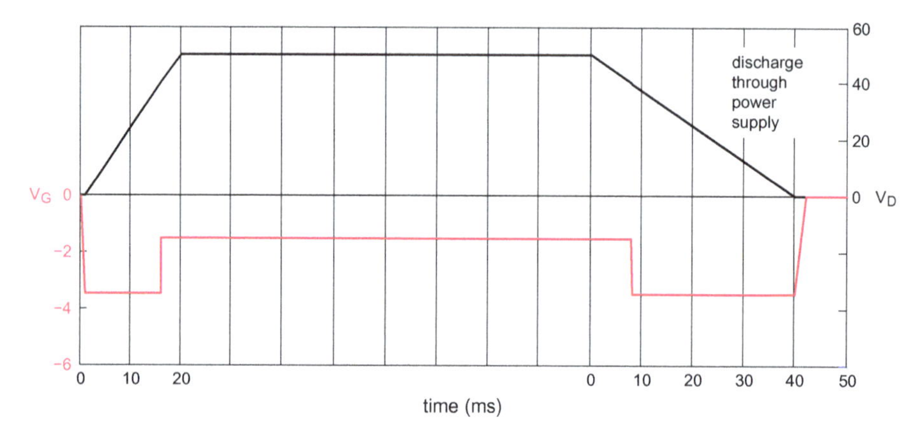

Fig. 2.15 Recommended gate and drain voltage sequence for GaN HEMT

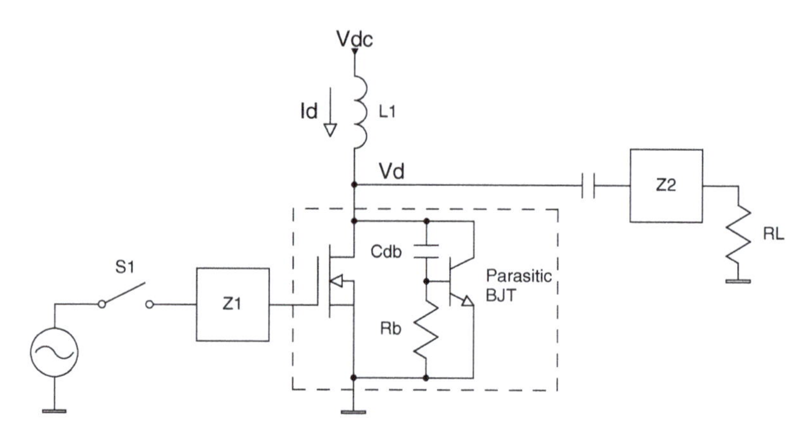

Fig. 2.16 LDMOS power amplifier with pulsed RF signal, showing parasitic BJT subject to snapback

2.3.6 *Pulse Considerations*

As discussed earlier (Sect. 2.2.4), LDMOS transistors include a parasitic BJT which can be triggered into destructive snapback under conditions of saturated operation and high $\mathrm{d}V_d/\mathrm{d}t$. When designing an LDMOS power amplifier which is used with pulsed (or other fast edge rate) signals, care must be taken not to trigger snapback.

Consider Fig. 2.16 which illustrates a class B LDMOS power amplifier with input and output matching networks Z_1 and Z_2, DC feed inductor L_1, and switch S_1 to pulse the RF input signal. The transistor is being operated at saturation, so V_d is a sinusoid with a peak voltage of V_{dc} swinging between 0 V and $2 \cdot V_{dc}$. Because I_{dc} is flowing through L_1, energy is stored in its magnetic field. When S_1 is opened, I_{dc} drops abruptly to 0 A, and the energy stored in L_1 is released. Because there is now no current flowing through the transistor, a transient voltage $V_d = V_{dc} + L_1 \frac{di_{dc}}{dt}$ is generated, where t is the time it takes for S_1 to open.

Energy is *also* stored in resonant circuit Z_2, which will continue to 'ring' and deliver power to load R_L *and* superimpose an exponentially decaying sinusoid on V_d. Thus until the energy stored in L_1 and Z_2 is dissipated in R_L and circuit losses, peak voltages as high as $2 \cdot V_{dc} + L_1 \frac{di_{dc}}{dt}$ can be present on the LDMOS drain.

Assume that the amplifier is delivering 500 W of power at η = 60% with V_{dc} = 32 V, so I_{dc} = 26 A. L_1 is a microstrip line of w = 1.6 mm, h = 0.8 mm, l = 36 mm, with calculated inductance = 10.7 nH [23]. If S_1 switches off in 10 ns, the peak drain voltage will be ≈ 92V. With a snapback threshold of 80 V, this will almost certainly trigger the parasitic BJT. What about a 20 ns transition time? That results in a 78 V peak voltage, which is uncomfortably close. Better to use an RF switch with a transition time of at least 30 ns and have some margin.

2.3.7 *Power Monitoring*

It is advisable to integrate DC current and RF power monitoring into every PA module to support system calibration and self-test, and in some generator architectures, for closed-loop power levelling. Smaller generators with a single PA module (e.g. Fig. 2.19) generally also need a dual directional coupler at the generator output (after any isolator) for measurement of reflected power. Given the high-power level, a coupling ratio of at least 30 dB is usually required. However, at our frequencies of interest coplanar microstrip coupled lines do not have enough isolation for a usable 30 dB directional coupler, so other methods must be used (Fig. 2.17).

2.3.8 *Integrated Power Amplifier Devices*

Because of the expertise and time needed to develop a reliable, manufacturable RF power amplifier, it would often be desirable to be able to purchase an off-the-

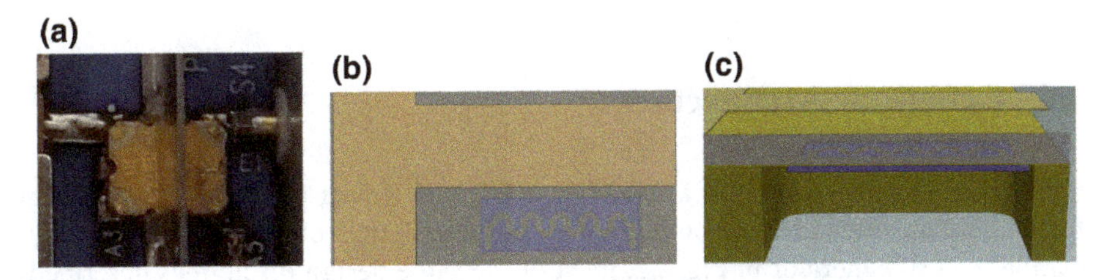

Fig. 2.17 Directional couplers for use in power amplifier PCBs: **a** Commercial 300 W, 2.45 GHz, 30 dB SMT coupler; **b, c** 1.6 kW, 2.45 GHz, 30 dB embedded coupler in 4-layer PCB. The coupled line is a stripline on layer 3 coupled to the main line on layer 1 through an offset aperture in the layer 2 ground plane. This provides excellent isolation and resulting 26 dB directivity

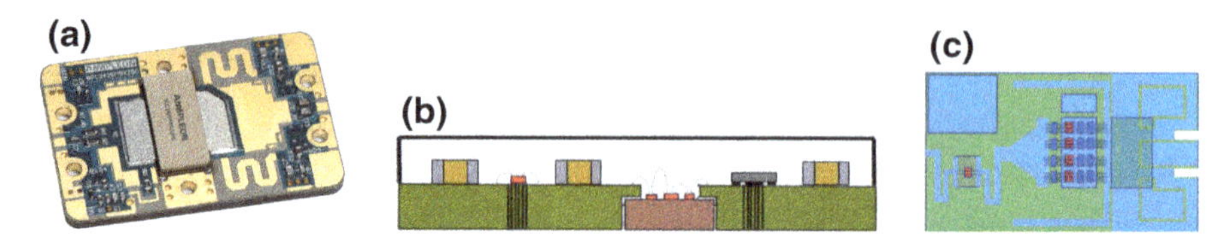

Fig. 2.18 Integrated power amplifier technologies: **a** pallet; **b**, **c** HPMCM structure. © Ampleon

shelf PA component with integrated matching networks and bias circuitry, just as RF integrated circuits now contain entire small-signal subsystems. Because high-power passive components (capacitors, inductors, transmission lines) are needed for impedance matching, it isn't possible to integrate the entire PA onto an IC die. However, RF power semiconductor manufacturers have been developing packaging and manufacturing technology to allow them to produce integrated PA devices at lower cost and/or smaller size than is possible by the generator manufacturer:

- Pallets, which use conventional multilayer PCB technology and microstrip matching networks, but are optimised for thermal performance and reproducibility, and are manufactured in a dedicated high-volume production process;
- High-power multichip modules (HPMCM), which use multilayer ceramic or high-Dk laminates with embedded copper coins, bare transistor die, and matching networks implemented with wire bonds, discrete MLCC capacitors, and integrated LTCC networks, enclosed in an air-cavity plastic package. Since the assembly process is an extension of that used for RF power transistors, those proven high-volume processes can be extended to HPMCMs as well.

These two technologies are illustrated in Fig. 2.18.

At least one of the major manufacturers (Ampleon) builds a family of standard 1- and 2-stage power amplifier pallets for the 433 MHz, 915 MHz, and 2.45 GHz ISM bands, with power levels from 60–700 W. Since these have 50 Ω inputs and outputs and are sold as fully tested and warrantied components, they eliminate the need for a generator designer to also design and manufacture the power amplifiers. In many applications, they can provide cost-effective and reliable building blocks for a complete generator, e.g. Fig. 2.19.

2.4 Generator Architecture

Figure 2.19 is a photograph of a 250 W, 2.45 GHz RF power generator. Despite the fact that it has only 0.3% the output power and 0.01% the volume of the massively parallel 75 kW generator in Fig. 2.2, most of the same design disciplines described in this chapter (with the notable exception of power combining) were applied to it as well.

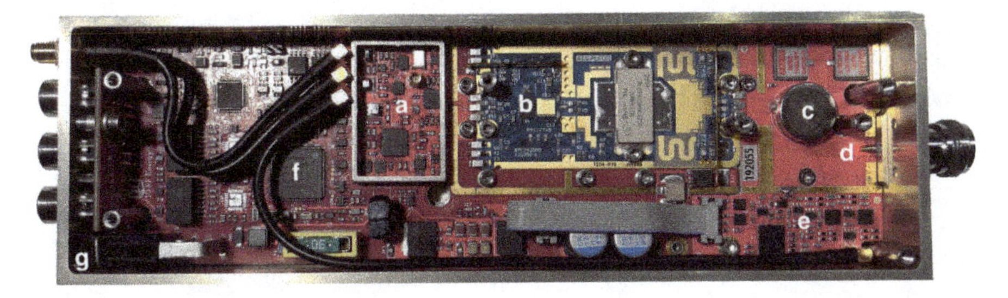

Fig. 2.19 Commercial 250 W 2.45 GHz generator (shields and cover removed): **a** frequency reference, synthesised RF source, and 1 W predriver; **b** 2-stage 300 W PA pallet; **c** circulator and isolation resistor; **d** embedded directional coupler; **e** power detectors and vector receiver; **f** MCU and communication interfaces; **g** 22 × 7 × 3 cm machined enclosure. © 3D RF Energy Corp

2.4.1 Gain Budgeting

A critical process in the design of any RF power generator that is frequently neglected or inadequately performed by inexperienced designers is system gain (loss) budgeting. This needs to be done in connection with thermal budgeting (Sect. 2.4.6) and updated over the course of system design and prototype testing. It can be structured in various ways, e.g. the budget for each subsystem is usually made part of the associated specification and delegated to the team responsible for designing that subsystem. However it is managed, every part of the generator system must be considered.

Starting at the top, what is the generator specification—minimum power level, over what frequency range, over what ambient temperature range? Given that the gains of the power amplifiers and drivers are strong functions of temperature *and* frequency, and these effects are cascaded, the range of variation in system gain is often surprisingly high and must be considered from the outset.

- What is the worst-case loss in the output combiner (specified by the combiner manufacturer or determined by component-sensitivity analysis of the design)?
- What about the losses in the cables and adapters used to connect the PA modules to the combiner?
- What is the PA module's range of RF performance—
  - *Minimum* gain over frequency and its budgeted temperature range?
  - *Maximum* gain over frequency and temperature?
  - Minimum compressed power over frequency and temperature?
  - Range of gain and phase variation for a given frequency and temperature?
  - These should all be determined by component-sensitivity (e.g. Monte Carlo) analysis of the PA design. It isn't necessary to use the worst-case gain and power specified for the RF power transistor, but the analysis must define production test limits for the PA that will be sufficient to guarantee system performance.
- What are the worst-case losses in the power splitter and associated cables?

- What are the minimum and maximum gains and minimum *uncompressed* power of the driver stages over frequency and temperature?
- What is the power variation of the signal source over frequency and temperature?

It would not be surprising for the cascaded source, driver, and PA chain in a 2.4–2.5 GHz generator to exhibit 10 dB variation over a 100° ambient temperature range, plus 6 dB variation over frequency, plus 4 dB manufacturing and combining variations, resulting in a gain variation 20 dB greater than 'nominal', so

- The variable gain element used to adjust power level must have enough range to cover the sum of the
  - generator's specified power range;
  - power variation of the source;
  - range of driver gain from lowest-gain frequency at the highest temperature, to the highest-gain frequency at lowest temperature;
  - range of PA gain from lowest-gain frequency at the highest temperature, to the highest-gain frequency at lowest temperature;
  - range of PA gain from minimum required output power (where the gain is lowest), to high (but uncompressed) power;
  - range of loss in PA combining due to gain and phase mismatches (Sect. 2.4.3.8);
  - range of loss in combiner.
- The RF switch used for pulsed operation must have a minimum isolation (i.e. on/off ratio) equal to this range *plus* the generator's specified pulse on/off ratio, and the PCB layout must support the required isolation.
- EMI shielding and filtering must provide enough isolation between various parts of the system to prevent feedback-induced oscillation or instability at the highest-gain frequency at the lowest temperature.

2.4.2 RF Power Oscillators

'Why not just make a high-power oscillator?' is a question that probably has occurred to every engineer tasked with the design of a compact RF power generator for cooking. After all, that would be analogous to the magnetron, which is simply a power oscillator.

Since interest in solid-state cooking began, there have several well-documented power oscillators for this application, from a 310 W, 46% LDMOS PA [24] with PCB resonator to an injection-locked 210 W, 51% LDMOS PA [25]. However, all power oscillators suffer from some common drawbacks in comparison with a conventional signal source and amplifier lineup:

- negligible improvement in system efficiency;
- negligible reduction in system size;
- limited range of power control (typically < 3 dB);

- slow on/off switching time (typically 10 ms).

In the end, the size, efficiency, and cost of a single-output generator are dominated by the PA—which is the same for a power oscillator as for a conventional lineup.

2.4.3 *Power Combining*

Because most industrial and scientific RF power applications need more than the 200–700 W that can be generated by a single solid-state power amplifier, one or more stages of power combining are usually needed. Consistent with architectures successfully used in high-power solid-state television transmitters and particle accelerators, power is combined in three stages:

1. 2–6 individual power amplifiers are combined using a planar (PCB) combiner to create a PA module with a total power of 0.5–3 kW (e.g. Fig. 2.20). For modularity and ease of test, the PA module output is usually a coaxial connector.
2. 2–8 PA modules are combined into a waveguide output using structures such as radial combiners, for a total power up to 100 kW. With SIW technology (which readily accommodates transitions to both microstrip and waveguide ports), these two stages can be consolidated to deliver up to 8 kW in a single planar structure.
3. Higher power levels are achieved with waveguide combiners.

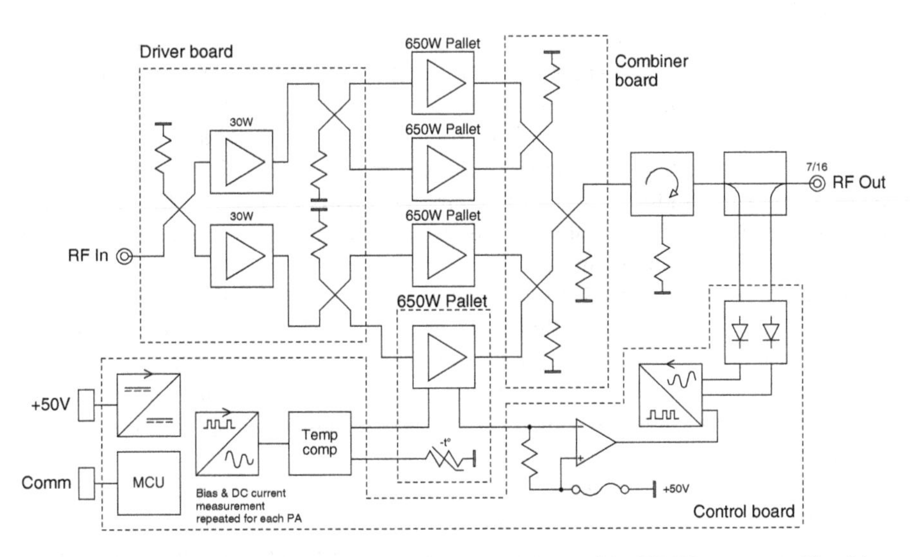

Fig. 2.20 Block diagram of 2.5 kW, 915 MHz PA module used in 75 kW generator. The driver uses SMD 3 dB couplers as the power splitter, and the combiner is fabricated as a PCB structure (Fig. 2.23). The control board includes temperature-compensated bias sources (Sect. 2.3.5.1) and RF and DC power monitors (Sect. 2.3.7)

2.4.3.1 Power Amplifier Isolation

An important consideration in any power combiner design is to ensure that—irrespective of isolation from the load—the outputs of individual power amplifiers are isolated from each other.

Recall that solid-state PAs do not present a constant output impedance, but rather are designed to deliver a specific power into a given load at maximum efficiency. As a consequence, the non-constant output impedance of a second PA connected to another can modulate the load seen by the first, both in magnitude and phase. We use this behaviour to our advantage in Doherty amplifiers: however, when coherently combining the power of multiple amplifiers, it works against us. At minimum, it makes it difficult to balance the power delivered by the individual amplifiers: most often, it creates RF oscillation in one or more amplifiers, or a lower frequency system oscillation as the various control loops (e.g. RF power levelling and DC power supply voltage sense) attempt to correct for changing power amplifier load conditions and consequent changes in DC current and delivered RF power. This system oscillation manifests as undesired low-frequency amplitude modulation of the output power. Some combiner structures (e.g. quadrature and chain combiners) intrinsically provide this isolation; with others (e.g. reactive planar and radial combiners), the isolation must be provided by circulators.

2.4.3.2 Reactive Combiners

The simplest power combiner is simply connecting N in-phase sources together at a single point into an output load $R_{\text{out}} = R_{\text{in}}/N$. Unfortunately for PCB combiners, the required load impedance R_L is usually the same as R_{in} (i.e. $50\,\Omega$), so a transformer is needed from R_{out} to R_L. Since it is impractical to fabricate a wide impedance range of power-handling microstrip on a given substrate—$12.5\,\Omega$ microstrip is more than six times the width of $50\,\Omega$, for example—this is impractical for $N > 4$. For this reason, reactive PCB combiners are usually constructed as a 'tree' of 2-way or 3-way combiners, as shown in Fig. 2.21.

We have already discussed the importance of providing isolation (i.e. circulators) between individual power amplifiers and a reactive combiner. This leads designers to conclude that no additional isolation is needed between the combiner output and the final load. Indeed, provided that the circulators at each combiner input are adequately sized, they will protect the individual power amplifiers from damage caused by power reflected from a mismatched load.

However, the entire combiner structure is subject to the reflected power, and at high VSWR the current standing wave will exhibit current maxima as much as twice that under matched conditions. This subjects the conductors in areas of higher current density (e.g. the corners of microstrip mitred lines and junctions, as illustrated in Fig. 2.22) to *four* times normal power dissipation. Unless this is taken into account in the combiner design, large load mismatches are likely to result in rapid cascaded failure of an entire section of the combiner.

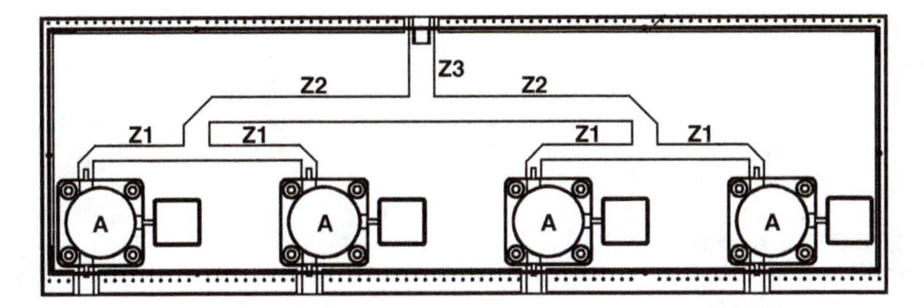

Fig. 2.21 4-way 1 kW combiner fabricated on 1.6 mm RT/duroid 6035HTC. Insertion loss (excluding circulators) is <0.2 dB at 2.45 GHz. **a** Circulators A are used to isolate each input port; **b** Lines Z_1 are arbitrary (but equal) lengths of 50 Ω; **c** Lines Z_2 are $3\lambda/4 \times 35.4\,\Omega$ to transform the 25 Ω Z_1 summing points back to 50 Ω; **d** The impedance at the end of the Z_2 transformers are once again 50 Ω, summing to 25 Ω, so line Z_3 is $\lambda/4 \times 35.4\,\Omega$ to transform back to 50 Ω at the output

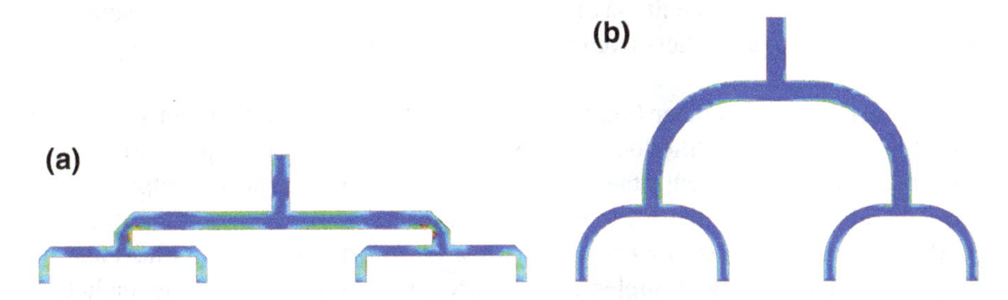

Fig. 2.22 Current density in 4-way microstrip coupler. **a** High current density at mitred microstrip corners creates hot spots and combiner failure under high VSWR. **b** Using radiused corners reduces the peak power density to 1/4 that with mitred corners. We can use 'S-curves' to reduce the total length of the combiner, but clearly it will be difficult to keep it as compact as the one with mitred corners

2.4.3.3 Quadrature Combiners

The quadrature (or hybrid) is a 4-port coupler in which there is a 90° phase shift between adjacent ports. If a signal is incident on one port only, there will be an 'isolated port' from which no power will exit, and the power will be divided (usually equally, although quadrature couplers can be designed for unequal power division) between the two non-isolated ports, but separated in phase by 90°. Whether used as a power divider or combiner, the isolated port is terminated in Z_0.

The quadrature coupler is by far the most popular method of combining two power amplifiers (the combined PA is sometimes called a 'balanced amplifier').

When used as a power divider, the quadrature coupler provides the advantage of isolating the input port from mismatched output ports, so long as both ports have the same reflection coefficient. The reflected power is dissipated in the load resistor terminating the isolated port. Since power amplifiers often have poor input return loss—but the phase and magnitude of the reflection coefficient tends to be quite

Fig. 2.23 4-way quadrature coupler used in 2.5 kW, 915 MHz PA module shown in Fig. 2.20. This is implemented as three cascaded branchline couplers fabricated on a 1.6 mm RO6035HTC PCB substrate with 800 W loads bolted to the cold plate as isolation resistors

reproducible from unit to unit—a quadrature coupler prevents the input mismatches of combined power amplifiers from modulating the load seen by their driving amplifier.

When used as a combiner (i.e. the isolated port is now on the output side of the coupler), the power amplifier outputs are isolated from each other, preventing them from load-modulating each other. Note that reflected power at the output port is divided and returned to the PA outputs, not absorbed in the isolation resistor. The quadrature combiner has low loss (because of the short transmission lines) and its simplicity makes it easy to implement in every type of transmission line, including microstrip and waveguide. Figure 2.23 shows a typical implementation.

2.4.3.4 Chain Combiners

The chain (also called serial or progressive) combiner shown in Fig. 2.24 is a coupler structure first proposed in 1961 [26]. Essentially, it uses N-1 directional couplers to add 1/N of the output power from each successive stage; thus, each successive stage has a higher coupling coefficient of $10\log_{10}N$ dB (neglecting losses).

The chain combiner has the advantages of being nonbinary and having the shortest line lengths between the sources (power amplifiers) and the combiner structure. In principle, any number of inputs could be combined. However, losses in the coupler reduce the combining efficiency and it becomes increasingly difficult to fabricate the coupler as the coupling coefficient increases. In practice, it is used successfully in solid-state UHF television transmitters with $N \leq 8$, and in 2.45 GHz ISM generators with $N \leq 6$. It is important to maintain low losses and low radiation in the couplers: for applications above 1 GHz, suspended stripline has proven to be a good technology for PCB implementations [27].

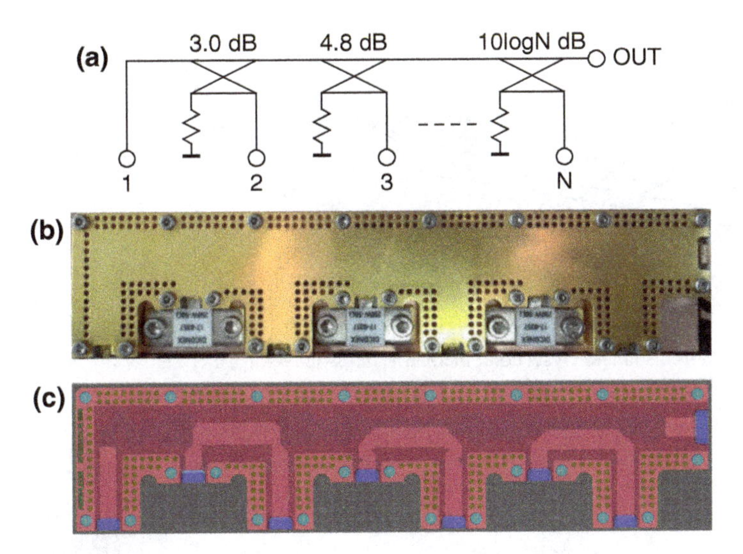

Fig. 2.24 Chain combiner: **a** schematic; **b** photograph of 4-way 2.45 GHz 1.2 kW combiner fabricated in suspended substrate with flange-mounted isolation resistors; **c** transparent view showing coupled inner layers. © Ampleon

2.4.3.5 Wilkinson and Gysel N-Way Combiners

The N-way combiner described in 1960 by Wilkinson [28] is well-known by RF designers, particularly the 2-way variant. It has the advantages of simplicity, low insertion loss, and isolation between all input ports. Its main disadvantage (and the reason the 2-way variant predominates) is that the isolation resistors from each input port need to be connected at a common node.[10] This becomes impractical for $N > 2$ in planar (PCB) structures.

In 1975, Gysel extended Wilkinson's design to a new N-way structure intended for high-power applications (Fig. 2.25) [29]. Although added transmission lines make this structure more complex than Wilkinson's, it is more practical for high-power applications because the isolation resistors are all one-port (grounded) Z_0 terminations. Gysel combiners quickly became very popular for kW-level power combining in broadcast transmitters [30], where they are typically implemented using rigid air-core coaxial transmission lines [31].

The port isolation and scalability of the Gysel topology make it attractive for high-power PCB combiners. Unfortunately, Wilkinson and Gysel combiners have a common characteristic that complicates high-power microstrip and stripline implementations: the $\lambda/4$ transmission lines between the input ports and the output have a characteristic impedance that increases with N (typically $\sqrt{N R_{in} R_{out}}$) [32], so they quickly become too narrow to handle high currents. Transmission line impedance

[10] Of course this can also be realised with an isolation resistor connected between each input port. But for $N > 3$, this gets ugly very quickly.

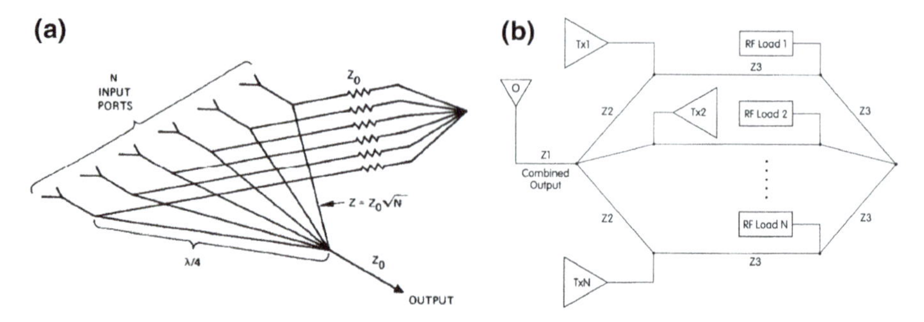

Fig. 2.25 Wilkinson and Gysel combiners. **a** Wilkinson N-way; **b** Gysel N-way

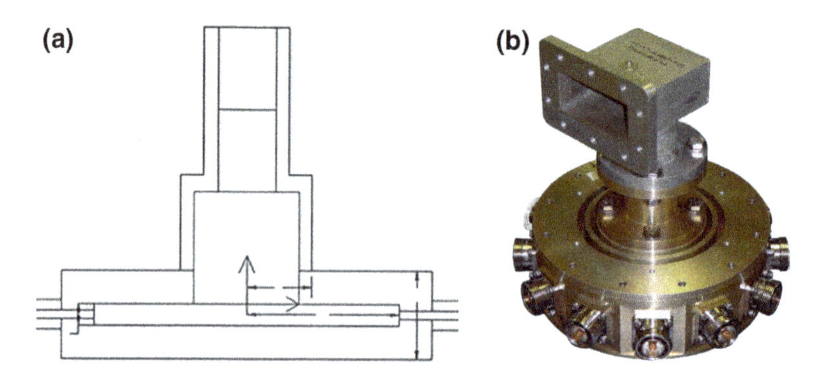

Fig. 2.26 Radial power combiners. **a** Structure of parallel plate radial transmission line combiner [35]; **b** Commercial 7 kW, 12-input combiner © ORBIT/FR

$> Z_{in}$ can be avoided by adding a $\lambda/4$ transformer between the summing node and the output port or by a variety of multi-stage topologies, and novel implementations such as back-to-back microstrip [33] and SIW-microstrip hybrids [34] demonstrate other possible approaches for constructing high-power Gysel combiners in PCB. Nevertheless, 2^N and N-way in-phase reactive (with per-channel isolators), 2^N quadrature, and N-way chain combiners continue to be the most practical structures for implementing 0.5–3 kW combiners in PCB.

2.4.3.6 Radial N-Way Combiners

Radial combiners are used for multi-kW combining in the 915 MHz and 2.45 GHz ISM bands. Although there are a variety of structures used for radial combiners (cavity vs. non-cavity, radial versus conical transmission lines), those used for ISM applications are generally N-way dielectric non-resonant radial transmission line structures, with inputs equally spaced around the circumference of a parallel plate radial transmission line, and output power extracted at the centre (Fig. 2.26) [35].

For solid-state generator applications, radial combiners are usually constructed with coaxial input ports (to accommodate the outputs of 1–3 kW RF power amplifiers) and a waveguide output port. These radial combiners usually do not provide isolation between input ports, so circulators are generally needed at the PA outputs. Radial combiners for these ISM applications typically provide 4–32 input ports with total average power handling up to 100 kW. Radial combiners' N-way single-step combining geometry and small size result in very low combining losses (typically <0.3 dB at 2.45 GHz), and their machined metal structures provide excellent shielding and durability.

2.4.3.7 Substrate Integrated Waveguide

Substrate integrated waveguide (SIW; Fig. 2.27), also called post-wall waveguide, is a new type of planar transmission line [36] which offers a number of benefits for RF power combiners and couplers above 1 GHz. It implements a waveguide within the dielectric substrate of a PCB by constraining the wave between two rows of posts or plated-through holes at the sides and ground planes above and below.

Because the sidewalls are not continuous, transverse magnetic (TM) modes do not exist, and TE_{10} is the dominant mode. Thus, the thickness of the substrate does not affect the cut-off frequency of the waveguide, but only the dielectric loss (a thicker substrate has lower dielectric loss). Let a and b denote the width and height of a rectangular waveguide. Assuming $a > b$, the propagating mode with the lowest cut-off frequency is TE_{10} (dominant mode). Using [37], we can calculate $a = \frac{c}{2f_c\sqrt{\varepsilon_r}}$, where f_c is the cut-off frequency of the TE_{10} mode, and c is the speed of light in free space.

So commercial microwave PCB substrates (e.g. Rogers RO3010, with a dielectric constant of 10.2) can be used to construct SIWs for the 2.45 GHz ISM band as narrow as 25 mm[11] [38]. The half mode substrate integrated waveguide (HMSIW) variant (which simply bisects the SIW structure transversely using a fictitious and symmetrical magnetic wall) can decrease this size by nearly 50% [39].

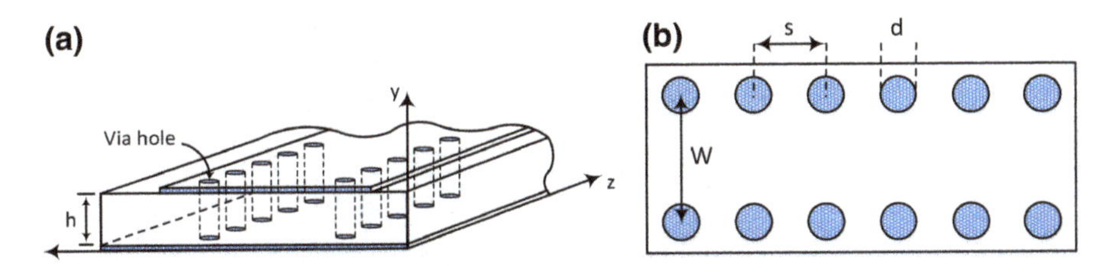

Fig. 2.27 Substrate integrated waveguide: **a** structure; **b** top view

[11] Since SIW sidewalls are not continuous, the effective width is slightly smaller than a_d. The most popular calculation of effective width a_e is the one proposed in [38], $a_e = a_d - d^2/0.95s$, where d and s are the diameter and spacing of the metal posts (all dimensions are in mm).

SIW inherits microstrip's ease of PCB integration while providing most of waveguide's benefits, including higher power capacity, high Q, and low radiation loss. Integration of many SIW circuits (e.g. combiners, couplers, and filters) is possible on the same PCB, and transitions to microstrip, stripline, and conventional waveguide are all straightforward. It is also practical to implement multi-port combiners with inherent bandpass filter characteristics in standard SIW technology.

Almost all H-plane waveguide combiners and couplers have been translated to SIW, and new topologies that make use of novel hybrid structures (e.g. SIW and microstrip [34]) are being developed (Fig. 2.28). As of 2019, the highest power combiners use 2-way (non-isolating) in-phase structures and (isolating) quadrature couplers, so are limited to combining 2^N sources. In simple SIW, directional couplers are limited to about 20 dB coupling ratio because of the practical limit of isolation in 2-layer SIW structures is 30–40 dB. However, multilayer SIW technologies [40] allow more complex structures and increase the possible isolation between different parts of

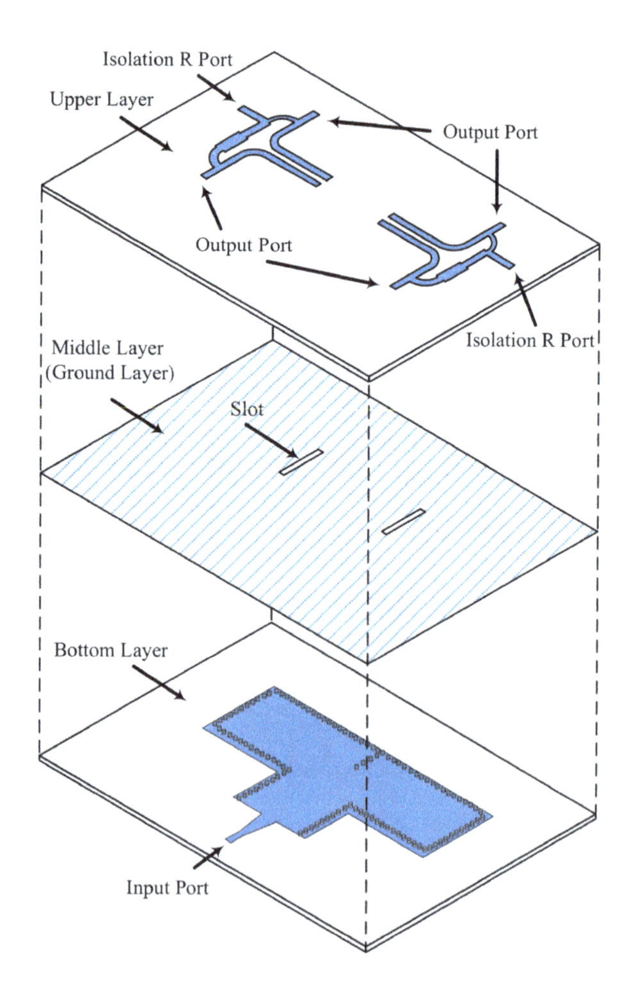

Fig. 2.28 SIW 4-way Gysel combiner [42] fabricated in a 3-layer PCB (RO4003C) structure. The test circuit was designed for operation at 8.4–10.6 GHz, but it is straightforward to scale to other frequencies

Table 2.2 Effects of phase mismatch for four combined 270 W, 68% efficient power amplifiers

Phase variation	Output power (dB)	Device dissipation (W)	Isolation dissipation (W)	System efficiency (%)
±10°	−0.04	1.2	10.5	−0.65
±15°	−0.09	2.7	22.6	−1.39
±20°	−0.16	4.8	40.8	−2.47

the circuit. Finally, it is feasible with SIW to implement high-power circulators [41], although integrating the required multi-kW isolation resistors remains a challenge.

2.4.3.8 Phase and Gain Matching

When multiple power amplifiers are combined, the total power will be reduced by losses in the combiner—and structures such as the radial combiner and SIW are used to reduce this loss as much as possible. However, phase difference between the various inputs to the combiner *also* introduces loss. Phase differences between amplifier channels can be the result of PA-to-PA manufacturing variations, phase differences between outputs of the splitter, or differences in the electrical length of interconnecting microstrip lines or coaxial cables.

Table 2.2 shows the effects of phase mismatch in a generator combining four PAs, each delivering 270 W output power at 68% efficiency, where *output power* indicates the change in total output power; *device dissipation* indicates the increased dissipation in each transistor; *isolation dissipation* indicates the total dissipation in all isolation loads; and *system efficiency* indicates the change in total system efficiency.

2.4.4 Signal Sources

Just as the cellular and wireless networking industries have done for RF power transistors, they have driven the development of high-performance RF integrated circuit and passive component technologies and led to a proliferation of standard parts. When designing a signal source for an RF power generator, the most common starting point is with a standard general-purpose PLL frequency synthesiser IC (Analog Devices and Texas Instruments lead this market). These commonly include an integrated VCO and an I/Q modulator, which can be used to implement power and phase control. The output power of these ICs is typically limited to about +10 dBm.

Several manufacturers have introduced fully integrated signal generator ICs for ISM power generators. One example (Ampleon BLP25RFE001) delivers +24 dBm of output power in the 433 MHz, 915 MHz, and 2.45 GHz ISM bands and includes gain and phase control and an RF switch.

For rapid prototype assembly and low-volume production, PCB-mount frequency synthesiser modules are a simple solution. These are available from a wide range of manufacturers, including API, EM Research, Luff Research, Mini-Circuits, Synergy Microwave, and Z-Comm.

All of these sources will need to be followed by one or more stages of driver amplifiers to develop the power needed to drive the final power amplifiers. A wide range of amplifiers with power levels up to about 1 W are available as 50 Ω gain blocks. For higher power levels, we generally use transistors, monolithic microwave integrated circuit (MMIC), and pallets from the same manufacturers who supply the LDMOS or GaN power transistors for the PA.

2.4.5 Coherent Measurements

Coherence is a valuable technique enabled by solid-state RF sources that is extensively used in RF test instruments to increase the speed and accuracy of measurements. We have employed it in RF power generators in two different ways.

2.4.5.1 Impedance Measurements

In certain plasma generation applications [43], a key element in the generator's ability to monitor and optimise the plasma is its ability to measure large-signal impedance (not just forward and reflected power) in real time. As shown in Fig. 2.29, we can accomplish this by using the same technique as a vector network analyser (VNA).

The incident and reflected voltages at the generator output are sampled by a dual directional coupler incorporated into the structure of the output PCB (after the isolator). Each of these incident and reflected voltage samples is resolved by an I/Q demodulator into a pair of DC components representing the real and imaginary voltage. The two mixers in each demodulator are driven in quadrature by continuous 0° and 90° reference signals generated by the synthesiser.

These four DC samples are digitised by a four-channel A/D converter. In ref. [43], this has a 30 ns aperture time and a 1 Ms/s sampling rate, which allows it to measure the complex forward and reflected voltages in every microwave pulse in the burst. The demodulator and A/D gains are very stable over temperature, so these measurements are also used to stabilise power level.

For an ideal coupler, the reflection coefficient (and hence the impedance) of the load can be computed from sampled forward and reflected voltages V_f and V_r as

$$\Gamma = \frac{V_r}{V_f} \tag{2.1}$$

In practice, two effects degrade this ideal system. First, the terminating impedance at the sampled ports of the coupler cannot be expected to be exactly 50 Ω. Second

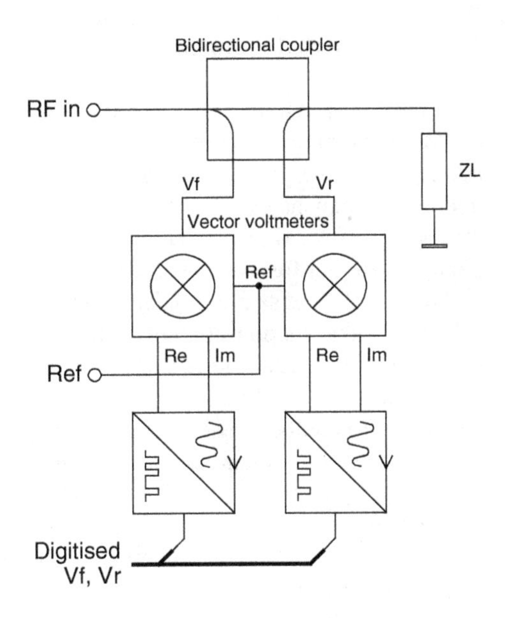

Fig. 2.29 Vector voltmeter used for real-time load impedance measurements

(and most troublesome), the coupler is non-ideal. However, we can show that by introducing three 'calibration constants' A, B, and C (complex numbers independent of Γ, but a function of frequency) we can relate the measured reflection coefficient Γ_M to the actual reflection coefficient as

$$\Gamma_A = \frac{\Gamma_M - B}{-C\Gamma_M + A} \tag{2.2}$$

To find constants A, B, and C, we measure the reflection coefficient (at every frequency of interest) with three standard devices (usually short, open, and load, each with accurately known Γ) connected to the output port, and solve the system of equations described in [44].

Of course, the load being measured is rarely at the generator output and is often on the other side of an impedance transforming element (e.g. applicator). However, if the connection between the generator and the load is represented as a 2-port S-parameter network, it can be mathematically 'de-embedded' [45] from Γ_A to determine the true reflection coefficient of the load Γ_L.

Although periodic recalibration can be impractical, often the only recalibration required is to periodically reestablish a phase reference for Γ, which is easily accomplished by measuring the reflection coefficient with no load present.

2.4.5.2 Coherent Sampling

In a multi-port system, it is often important to know (for chamber characterisation and/or load monitoring) how much power from a port is arriving at each of the other ports [46]. If the generators driving the other ports are operating at the same (or unknown) frequency, their power arriving at this port cannot easily be resolved from this port's reflected power. However, if the other generators are operating at even slightly different frequencies, their power can easily be resolved.

Fortunately, the sampled V_r data returned by the measurement system described in the previous section contains information not just of this generator's reflected power, but of all other power at a frequency within the A/D converter's bandwidth arriving at the port. And if we reference the I/Q demodulator not to the generator source but to another frequency synthesiser (locked to the same common reference clock), this measurement band can be shifted to whatever frequency we like.

For example, although in a 2.4–2.5 GHz system we could use an I/Q reference frequency of 2.4 GHz and an A/D sampling rate of 300 Ms/s to unambiguously digitise the entire ISM band, in practice we can use an inexpensive A/D converter with a 30 Ms/s sampling rate, and simply subsample the ISM band in ten segments by changing the I/Q reference frequency.

2.4.6 Thermal Management

Because RF performance and semiconductor lifetime are degraded by high operating temperatures, it is essential to consider thermal management from the outset in any power generator design. Figure 2.30 illustrates a simplified equivalent thermal circuit for a proposed air-cooled 1 kW generator which combines the outputs of five 250 W PA modules.

The PA is constructed as a PCB pallet with the LDMOS transistor soldered to an embedded 3 × 3 cm copper coin, so the thermal resistances of $R_{\theta cp}$ (solder) and $R_{\theta pp}$ (copper) are quite low. $R_{\theta ph}$ is the thermal interface between the pallet and the heatsink to which it is bolted and consists of a thin 50 μm layer of white silicone thermal grease (Fig. 2.31) [47]. $R_{\theta ha}$ is the thermal resistance of the heatsink, an 18 × 18 × 8 cm high-efficiency heatsink with 100 m^3/h airflow.

Although a transistor junction temperature of 137° is fine (this represents an MTF of 3 × 10^6 h), the 96° PCB surface temperature T_p may be a concern for other components (such as RF capacitors with self-heating), and their specifications should be reviewed. However, the primary contributor to heating is the heatsink: if air cooling is essential for this generator, a higher-efficiency heat exchanger should be considered, and the enclosure should be designed so that the small-signal and control electronics (which have lower operating temperature limits than the RF power transistors) can be cooled separately.

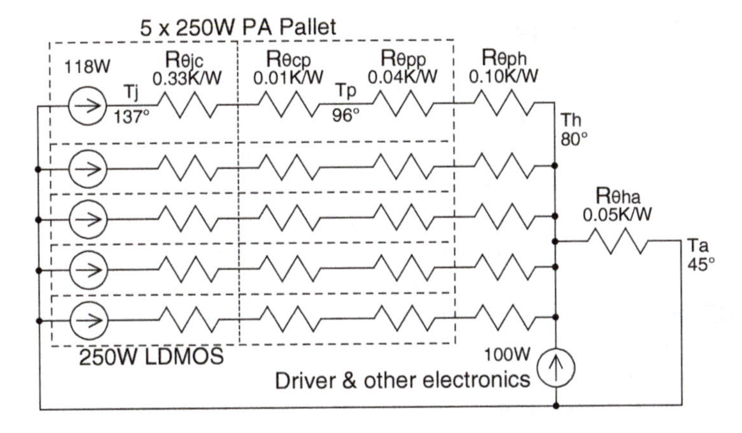

Fig. 2.30 Equivalent thermal circuit for 1 kW power generator using five 250 W PA pallets. $R_{\theta jc}$ is the thermal resistance of the transistor; $R_{\theta pp}$ is the resistance through the PA pallet baseplate; $R_{\theta cp}$ and $R_{\theta ph}$ are the resistances of the transistor–pallet and pallet–heatsink thermal interfaces, respectively; and $R_{\theta ha}$ is the heatsink–ambient resistance

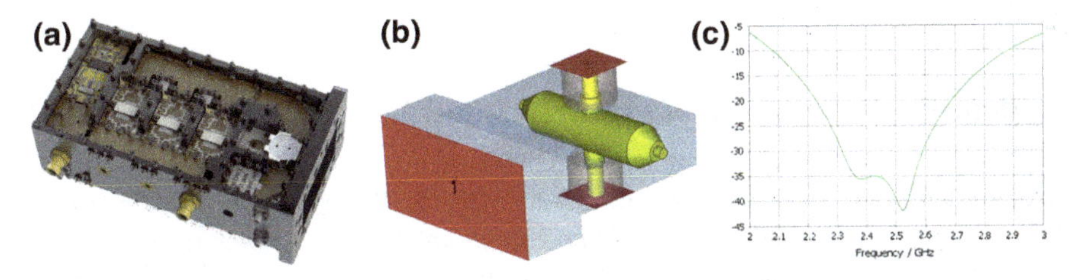

Fig. 2.31 Compact 1.4 kW, 2.45 GHz PA: **a** cover removed to show one of the 700 W PAs (each consisting of three 250 W PA pallets and an isolator) which are combined in a shared antenna to deliver 1.4 kW into the waveguide output; **b**, **c** EM simulation of the waveguide antenna

2.5 Design Tools

Although the Smith chart arguably remains the most important tool for designing microwave impedance matching networks, it is impractical to design an RF power generator without the aid of computer-aided simulation of circuits, PCB transmission lines structures, SIW and waveguide structures, and heat transfer.

ADS and Microwave Office (AWR) are the dominant commercial circuit and planar EM simulation tools used for RF power amplifier design. COMSOL, CST, HFSS, Sonnet, and QuickWave are popular commercial 3D EM simulators.

Although most of these tools are based on algorithms and libraries originally developed at universities, their owners have invested heavily in turning that foundation into powerful, easy-to-use tools; and when used for commercial product development, their high cost can be justified by the resulting saving in design time. Nevertheless, these capabilities can also be provided by open-source software, which provides benefits beyond the initial cost (zero) of the software—in particular, active communities that foster collaboration, open exchange, and a better fundamental understanding of the problem being solved.

Qucs is an open-source circuit simulator supporting S-parameter, AC, DC, transient, harmonic balance, and digital simulations [48]. It includes an optimiser, extensive libraries, and tools such as a transmission line calculator and filter synthesis application. It allows the creation of interactive nonlinear equation-defined device models based on Verilog-A [49]. It supports nonlinear models of LDMOS and GaN transistors; however, manufacturers only provide models for their devices as encrypted libraries for ADS and AWR, so those cannot be used in Qucs.

OpenEMS is an open-source 3D electromagnetic field solver using the FDTD (finite-difference time-domain) method in Cartesian and cylindrical coordinates [50, 51]. It includes many geometrical primitives and lumped circuit elements and can import from and export to mechanical CAD, PCB layout tools, and circuit simulators. It uses Octave or MATLAB as its scripting interface.

OpenFOAM is the leading open-source software for computational fluid dynamics (CFD) problems, most relevantly thermal simulation. It is heavily supported and is used as the computational engine for simulators such as simFlow.

Appendix: Calculation of Optimal Load Impedance

Readers with any exposure to circuit theory will recall the *maximum power transfer theorem*, which states that to obtain maximum external power from a source with a finite internal resistance R_S, the resistance of the load R_L must equal R_S as viewed from the output terminals. As extended to AC circuits, it states that maximum power transfer occurs when the complex load impedance Z_L is equal to the complex conjugate of the source impedance Z_S.

This is a perennial locus of misconception among new amplifier designers, who fail to appreciate that the theorem states *only* how to choose (for maximum power transfer) R_L *for a given* R_S: it does *not* say how to choose R_S for a given R_L. For a power generator, the objective is to deliver power to the load as *efficiently* as possible, and a simple thought experiment should convince the reader that in a purely resistive circuit, maximum power will be delivered to R_L (at 100% efficiency) when $R_S = 0$.

For the purpose, then, of designing a lossless impedance matching network between an RF power transistor and the load, it is more productive to turn this on its head and ask the question '*For a given amplifier class and supply voltage V_{dc}, what is the load R_L into which the transistor can most efficiently deliver power P_L?*.' Since amplifier classes are described in terms of their drain voltage and current, it is most instructive to start there.

For high-efficiency amplifier classes, the transistor is in its active region (i.e. conducting) for less than a complete cycle of input voltage V_{IN}. The circuit shown in (Fig. 2.32) includes an AC-coupled load and a conceptual harmonic short. All harmonics of the load are shorted and generate no voltage, so the drain voltage is a sinewave with a maximum amplitude of V_{dc}. Since $P = V_{(rms)}^2/R$, the maximum

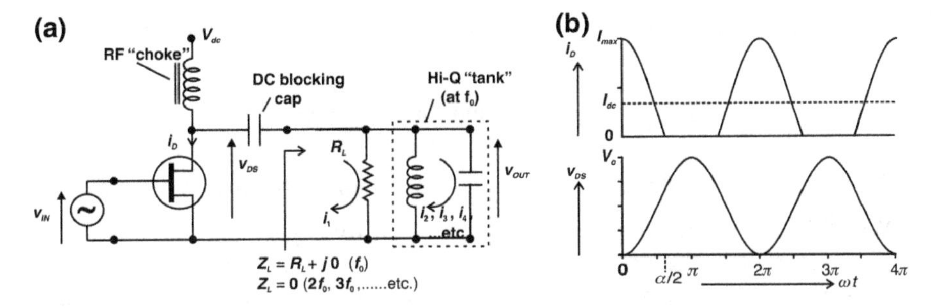

Fig. 2.32 Circuit and waveforms for ideal class B amplifier

power P_L that can be delivered to load R_L is

$$P_L = \frac{\left(\frac{V_{dc}}{\sqrt{2}}\right)^2}{R_L} = \frac{V_{dc}^2}{2R_L} \quad \text{and} \quad R_L = \frac{V_{dc}^2}{2P_L} \tag{2.3}$$

The mean (DC) current (including harmonics up to $5f_0$) is given by [52]

$$I_{dc} = \frac{I_{\max}}{2\pi} \cdot \frac{2\sin(\alpha/2) - \alpha\cos(\alpha/2)}{1 - \cos(\alpha/2)} = \frac{I_{\max}}{\pi} \quad \alpha = \pi \text{ (class B)} \tag{2.4}$$

and the fundamental component of the RF current waveform by

$$I_1 = \frac{I_{\max}}{2\pi} \cdot \frac{\alpha - \sin\alpha}{1 - \cos(\alpha/2)} = \frac{I_{\max}}{2} \quad \alpha = \pi \text{ (class B)} \tag{2.5}$$

The output ('drain') efficiency for a theoretical class B amplifier is then

$$\eta = \frac{P_1}{P_{dc}} = \frac{\frac{V_{dc}}{\sqrt{2}} \cdot \frac{I_1}{\sqrt{2}}}{V_{dc} I_{dc}} = \frac{I_1}{2I_{dc}} = \frac{\pi}{4} \quad (\approx 78.5\%) \tag{2.6}$$

This analysis can be extended to consider the effects of the knee voltage V_k described in Sect. 2.2.1.1, the drain voltage at which the maximum drain current is reduced to $(1 - e^{-1}) I_{\max}$, or about 63% of its saturated value at higher voltages. A normalised value of $V_k / V_{dc} = 0.1$ is a realistic value for modern LDMOS. We now need to reduce the voltage swing commensurately (i.e. by 10%) to prevent V_D from dropping below V_k and cutting off I_D, as this will significantly reduce power. With voltage swing reduced by 10%, R_L must also be reduced by 10% in order for I_D to reach $I_{\max}$. The result is that P_1 and η will both be reduced by about 10%. The effects of V_k are examined in detail in [53].

We can update our calculation of R_L (Eq. 2.3) to include the effects of V_k:

$$P_L = \frac{(V_{dc} - V_k)^2}{2R_L} \quad \text{and} \quad R_L = \frac{(V_{dc} - V_k)^2}{2P_L} \tag{2.7}$$

From the results of this exercise, we can make some important observations:

- With 28–50 V supply voltages, R_L is quite small for any significant power, e.g. a 32 V LDMOS device needs $R_L = 1.4\,\Omega$ to deliver 300 W. An internal matching network will be needed in this transistor to make final PCB matching practical.
- $R_L \propto V_{dc}^2$ (e.g. $R_L = 3.7\,\Omega$ for 300 W with a 50 V GaN device), so higher supply voltages significantly simplify the design of output matching networks and/or can eliminate internal matching networks altogether.
- For high-efficiency amplifiers, the whole concept of output impedance has broken down, since the waveforms are no longer sinusoidal. In the class B example just discussed, the device is completely cut off for half of the cycle, so the instantaneous output impedance takes on a switching characteristic. It is not possible to model the output of a high-efficiency PA with a Thevenin's equivalent.
- Because there is little scope to further reduce V_k in modern LDMOS, the practical high-power class B efficiency limit is likely about 74% for 50 V technology (915 MHz) and 72% for 30 V (2.45 GHz). Higher efficiencies than that will almost certainly be achievable only with switched-mode classes.

References

1. Collins GB (1948) Microwave magnetrons, vol 6. McGraw-Hill Book Company
2. Dexter A (2014) Phase locked magnetrons for accelerators. In: 27th international linear accelerator conference (LINAC14), pp 751–755
3. Wesson R (2016) RF solid state cooking white paper. Tech. rep, Ampleon
4. Kuo L, Kuo YH (1976) Chinese folk tales. Celestial Arts
5. Johnsen RJ, Granberg H (1979) Design, construction, and performance of high power RF VMOS devices. In: 1979 international electron devices meeting. IEEE, pp 93–96
6. Runton DW, Trabert B, Shealy JB, Vetury R (2013) History of GaN: High-power RF gallium nitride (GaN) from infancy to manufacturable process and beyond. IEEE Microw Mag 14(3):82–93
7. Theeuwen SJCH, Mollee H, Heeres R, Van Rijs F (2018) LDMOS technology for power amplifiers up to 12 GHz. In: 2018 13th European microwave integrated circuits conference (EuMIC). IEEE, pp 162–165
8. Lim TC, Armstrong GA (2006) The impact of the intrinsic and extrinsic resistances of double gate SOI on RF performance. Solid-State Electron 50(5):774–783
9. Van Rijs F, Theeuwen SJCH (2018) Efficiency improvement of LDMOS transistors for base stations: towards the theoretical limit. In: 2006 international electron devices meeting. IEEE, pp 1–4
10. Aadit MNA, Kirtania SG, Afrin F, Alam MK, Khosru QDM (2017) High electron mobility transistors: performance analysis, research trend and applications. In: Different types of field-effect transistors: theory and applications, chap. 3. Books on Demand, pp 45–64
11. Gurnett K, Adams T (2006) GaN makes inroads in the wireless infrastructure. III-Vs Rev 19(9):33–35

12. Ampleon (2013) Lifetime of BLF574XR in broadcast and ISM applications. Application Note AN11287
13. Black JR (1969) Electromigration failure modes in aluminum metallization for semiconductor devices. Proc IEEE 57(9):1587–1594
14. Burdeaux DC, Burger WR (2011) Intrinsic reliability of RF power LDMOS FETs. In: 2011 international reliability physics symposium. IEEE, pp 5A.2.1–5A.2.9
15. Curtice W, Pla J, Bridges D, Liang T, Shumate E (1999) A new dynamic electro-thermal nonlinear model for silicon RF LDMOS FETs. In: 1999 IEEE MTT-S international microwave symposium digest (Cat. No. 99CH36282), vol 2, IEEE, pp 419–422
16. Hammes P, Monsauret N, Loysel S, Schmidt-Szalowski M, van der Zanden J (2017) A robust, large-signal model for LDMOS RF power transistors. Microwaves & RF 56
17. Cripps SC (2006) RF power amplifiers for wireless communications. Artech House
18. Raab FH (2001) Class-E, class-C, and class-F power amplifiers based upon a finite number of harmonics. IEEE Trans Microw Theory Tech 49(8):1462–1468
19. NXP (2019) MRF24G300H RF power GaN transistor data sheet. https://www.nxp.com/docs/en/data-sheet/PRF24G300HS.pdf
20. Ampleon (2011) LDMOS bias module. Application report CA-330-11
21. Ampleon (2011) Bias module for 50 V GaN demonstration boards. Application note AN11130
22. Alim MA, Rezazadeh AA, Gaquiere C (2015) Thermal characterization of DC and small-signal parameters of 150 nm and 250 nm gate-length AlGaN/GaN HEMTs grown on a SiC substrate. Semicond Sci Technol 30(12):125005
23. Paul CR (2011) Inductance: loop and partial. Wiley
24. Shi T, Li K (2012) High power solid-state oscillator for microwave oven applications. In: 2012 IEEE/MTT-S international microwave symposium digest. IEEE, pp 1–3
25. Ikeda H, Itoh Y (2018) 2.4-GHz-band high-power and high-efficiency solid-state injection-locked oscillator. IEEE Trans Microw Theory Tech 66(7):3315–3322
26. Mohr R (1961) A microwave power divider (correspondence). IRE Trans Microw Theory Tech 9(6):573–573
27. Tahara Y, Oh-Hashi H, Ban T, Totani K, Miyazaki M (2001) A low-loss serial power combiner using novel suspended stripline couplers. In: 2001 IEEE MTT-S international microwave sympsoium digest (Cat. No. 01CH37157), vol 1, IEEE, pp 39–42
28. Wilkinson EJ (1960) An N-way hybrid power divider. IRE Trans Microw Theory Tech 8(1):116–118
29. Gysel UH (1975) A new N-way power divider/combiner suitable for high-power applications. In: 1975 IEEE-MTT-S international microwave symposium. IEEE, pp 116–118
30. Mendenhall GN, Shrestha M, Anthony E (1992) FM broadcast transmitters. In: NAB engineering handbook. Taylor & Francis, pp 561–562
31. Aves D, Kolvek SJ (1999) N-way RF power combiner/divider. U.S. Patent 5 880 648
32. Ardemagni F (1983) An optimized L-band eight-way Gysel power divider-combiner. IEEE Trans Microw Theory Tech 31(6):491–495
33. Beyragh DS, Abnavi S, Motahari SR (2010) Implementation of N-way Gysel combiners using back to back microstrip structure. In: 2010 IEEE international conference on ultra-wideband, vol 2. IEEE, pp 1–4
34. Chen H, Wang X, Che W, Zhou Y, Xue Q (2018) Development of compact HMSIW Gysel power dividers with microstrip isolation networks. IEEE Access 6:60429–60437
35. Jain A, Sharma D, Gupta A, Hannurkar P, Pathak S (2013) Compact solid state radio frequency amplifiers in kW regime for particle accelerator subsystems. Sadhana 38(4):667–678
36. Wu XH, Kishk AA (2010) Analysis and design of substrate integrated waveguide using efficient 2D hybrid method. Synth Lect Comput Electromagn 5(1):1–90
37. Pozar DM (2011) Microwave engineering. Wiley, p 113
38. Cassivi Y, Perregrini L, Arcioni P, Bressan M, Wu K, Conciauro G (2002) Dispersion characteristics of substrate integrated rectangular waveguide. IEEE Microw Wirel Compon Lett 12(9):333–335

39. Lakhdhar S, Harabi F, Gharsallah A (2017) Novel compact power dividers designs and comparison. In: 2017 international conference on engineering & MIS (ICEMIS). IEEE, pp 1–5
40. Djerafi T, Wu K (2016) Multilayer integration and packaging on substrate integrated waveguide for next generation wireless applications. In: 2016 46th European microwave conference (EuMC). IEEE, pp 858–861
41. Bochra R, Mohammed F, Tao J (2014) Analysis of S band substrate integrated waveguide power divider, circulator and coupler. Int J Comput Sci Eng Appl (IJCSEA) 4(2)
42. Moznebi AR, Afrooz K (2015) Four-way substrate integrated waveguide (SIW) power divider/combiner for high power applications. J Commun Eng 4(2):122–131
43. Williams R, Ikeda Y (2015) Real-time impedance measurement and frequency control in an automotive plasma ignition system. In: 2015 IEEE MTT-S international microwave symposium. IEEE, pp 1–4
44. Edwards ML (2001) Calibration and measurement of S-parameters. In: Microwave & RF circuits: analysis, design, fabrication, & measurement, chap. 7. Unpublished
45. Agilent (2004) De-embedding and embedding S-parameter networks using a vector network analyzer. Application note 1364-1
46. Yakovlev VV (2015) Frequency control over the heating patterns in a solid-state dual-source microwave oven. In: 2015 IEEE MTT-S international microwave symposium. IEEE, pp 1–4
47. Narumanchi S, Mihalic M, Kelly K, Eesley G (2008) Thermal interface materials for power electronics applications. In: 2008 11th intersociety conference on thermal and thermomechanical phenomena in electronic systems. IEEE, pp 395–404
48. Brinson ME, Jahn S (2009) Qucs: a GPL software package for circuit simulation, compact device modelling and circuit macromodelling from DC to RF and beyond. Int J Numer Model Electron Netw Devices Fields 22(4):297–319
49. Brinson ME, Kuznetsov V (2016) A new approach to compact semiconductor device modelling with Qucs Verilog-A analogue module synthesis. Int J Numer Model Electron Netw Devices Fields 29(6):1070–1088
50. Liebig T (2015) OpenEMS—open electromagnetic field solver. http://openEMS.de
51. Doñoro DG (2014) A new software suite for electromagnetics. Ph.D. thesis, Universidad Carlos III de Madrid
52. Cripps SC (2006) RF power amplifiers for wireless communications. Artech House, pp 40–43
53. Quaglia R, Shepphard DJ, Cripps S (2016) A reappraisal of optimum output matching conditions in microwave power transistors. IEEE Trans Microw Theory Tech 65(3):838–845

Part II
Heating Applications

Chapter 3
Mechanism of Microwave Heating of Matter

Noboru Yoshikawa

Abstract If we irradiate an object with microwave and observe its temperature increase, we would recognize the occurrence of "microwave heating." This article starts with discussing what is heat and what happens in matters during heating. And the interaction between microwave and the matter is learned from an atomistic point of view. Microwave is an electromagnetic wave and has components of alternating electric (*E*-) and magnetic (*H*-) field. It is pointed out that microwave *E*- and *H*-field interactions are different with materials. The materials are divided into three classes, and the heating mechanisms are discussed by each class of them. Heat is a state of either translational motion of molecules or vibration of lattice consisting of atoms (ions) or elevation of free electron energy in metals. Their moving frequency is much higher than microwave. Nevertheless, these high-frequency motions are excited by microwave of lower frequency. The mechanism of which is not necessarily understood well but is the "state-of-art" in understanding the microwave heating. This article discusses it and provides some clue for this interpretation.

3.1 What Is Heat?

If we irradiate an object with microwaves and observe its temperature increase, we would recognize the occurrence of "heating." However, it is not necessarily easy to measure the temperature in microwave heating because of restrictions in the measurement of the internal region of the objects without disturbing the microwave irradiation conditions; for example, the disturbance of the electric field distributions occurs by inserting a thermocouple. Here, it has to be kept in mind that all the microwave energy, which once absorbed by matter (heated objects), does not necessarily contribute to the increase of temperature or to heat matter. Microwave energy absorbed by matter could be re-emitted by some mechanisms and could be consumed for some structural changes that might take place in the crystalline structure of matter. These

N. Yoshikawa (✉)
Department of Materials Science and Engineering, Graduate School of Environmental Studies, Tohoku University, 6-6-02, aza-Aoba, Aramaki, Aoba-ku, Sendai 980-8579, Japan
e-mail: yoshin@material.tohoku.ac.jp

S. Horikoshi and N. Serpone (eds.), *RF Power Semiconductor Generator Application in Heating and Energy Utilization*, https://doi.org/10.1007/978-981-15-3548-2_3

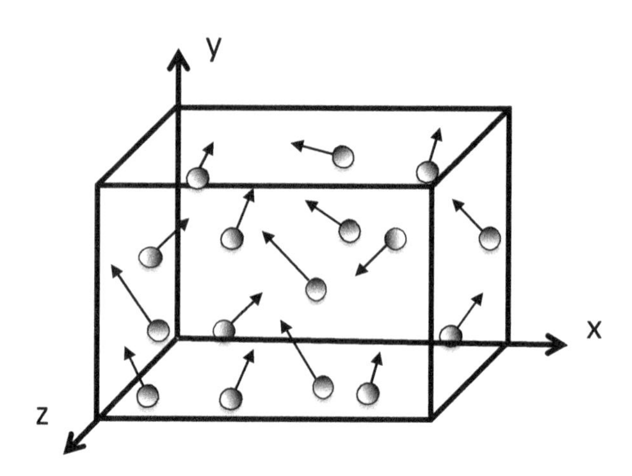

Fig. 3.1 Movement of gas molecules

phenomena are rather special and have thus far been discussed as the "non-thermal effect of microwaves." Accordingly, this chapter will deal mainly with the normal "microwave thermal effects" and "microwave heating."

Before discussing the microwave heating mechanism of matter [1–5], it is of importance to consider "what is heat." In the case of gases, we know of situations where gas molecules move at certain intermolecular distances, which are much larger than their molecular radii as schematically depicted in Fig. 3.1. They often collide with each other, the rates of which determine their viscosity and their heat conductivity. Increase of the average kinetic energy caused by the enhancement of the molecules' movement (velocity of molecules' movement) corresponds to the increase of the gas temperature and the kinetic energy, herein defined as heat (kinetic energy). This idea applies to water molecules in the gaseous phase (vapor), for example.

On the other hand, what happens if the kinetic energy decreased (temperature decrease) and the water vapor condensed into a liquid or even to ice (solid)? The temperature of liquid water may also be defined as the kinetic energy of the molecules although their moving range is much restricted, compared to the vapor state. In the solid state (ice), the molecules form a crystal, they can occupy certain positions of the lattice, and they can only move around these lattice positions (in a further restricted range). The displacement of the molecular position is determined by the superposition of lattice vibrational waves displaying some frequency spectrum. Nonetheless, heat is still defined as the kinetic energy of the molecules' movement. In fact, this definition can also apply to the case of organic molecules in the gas phase, in the state of liquid solvents and even in the crystalline state.

In the case of some inorganic solid substances, such as glasses used for windows and bottles, they may be amorphous and composed of positive and negative ions bound by ionic forces. They do not consist of molecules. In ceramic materials (either crystalline or amorphous), atoms are bound either by ionic or by covalent bonding, but they cannot be said to be molecules. Heat is then defined as the increase of the kinetic energy of the component atoms (ions).

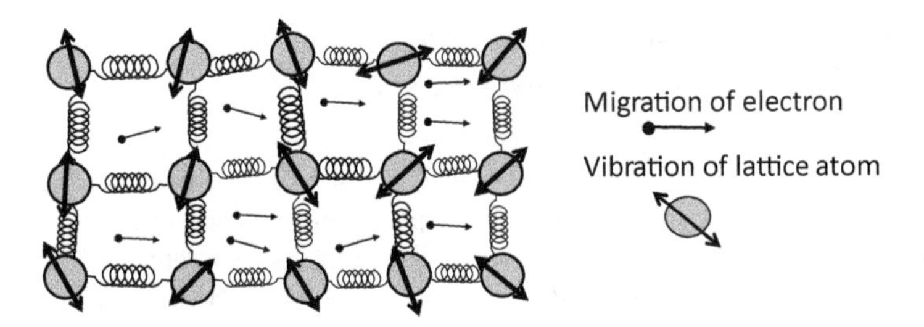

Fig. 3.2 Partitioning of kinetic energy to electron and lattice. Thermal vibration of metal crystal lattice

What about metals? States of metals can be either a solid, a molten state and even a vapor state. Most solid metals form crystalline states (some alloys become amorphous by rapid cooling). They do not consist of molecules either. A most different point is that metals have free electrons, which are not localized to a particular metal atom (positive ion) of the crystal lattice and thus are free to move around. A temperature increase in metals is not only accompanied by lattice vibrations but also by an increase in the kinetic energy of free electrons. In other words, the heat energy is divided into the kinetic energy of lattice vibrations of the positive ions and the energy of the free electrons. A schematic of a metal lattice is illustrated in Fig. 3.2, where ions are bound with springs (corresponding to the binding force) and they vibrate (the wavelength of which is larger than the interatomic distances), while the free electrons move through the clear spaces between the ions.

Summarizing the above arguments, heat means an increase in the kinetic energies of atoms, ions, molecules and electrons that make up matter. At this stage, however, in order to understand "the heat" in different materials, let us classify the kinetic energy (liquids, solids) of various materials as follows:

- **Class I**. Movement of molecules in liquids such as water or alcohol. The movement occurs in the molecular unit (e.g., rotation, translational motion, and vibrations within the molecules).
- **Class II**. (inorganic solids). Movement of atoms, ions or group of atoms which are either in the crystalline or in the glass (amorphous) states. Organic solid crystals or liquid crystals are also included.
- **Class III**. Movement of ions and free electrons in metallic materials (alloys, intermetallics, their molten states, crystalline or amorphous solids).

3.2 Difference Between Microwave Frequency and Vibrations of Atoms/Molecules

Microwaves consist of electromagnetic waves. When they propagate in free space, the electric and magnetic fields oscillate perpendicular to each other; the transmission direction is normal to both oscillating directions as illustrated in Fig. 3.3. The microwave frequency is of the order of 10^9 Hz (or 1 GHz). The 2.45-GHz frequency is adopted for domestic microwave ovens. The wavelength of 2.45-GHz microwaves is about 12 cm in air. Although the wavelength is diminished when the microwave penetrates through matter by an inverse square root of the relative permittivity (and the relative magnetic permeability), we may assume that the scale of the objects being heated (being heated in a microwave oven) is of similar order to that of the wavelength. Therefore, objects having dimension of half a wavelength experience the same direction of oscillating electric and/or magnetic field instantaneously. This situation is schematically illustrated in Fig. 3.4.

Microwave frequency corresponds to that of the rotational motion of water molecules. **Class I** substances such as water molecules in the liquid state have rotational, translational and stretching of atomic bonding motions. The motions become more active as the temperature increases. For example, rotational motion at 15 $\times$ 10^9 Hz corresponds to a temperature of 0.6 K, according to the equation $k_B T = h\nu$ (h: Plank constant, k_B: Boltzmann constant, T: Kelvin temperature, ν: frequency). A temperature of 50 °C (323 K)—maybe we feel "being heated"—corresponds to the translational motions of molecules, the frequency of which is of the order of 10 THz (10^{13} Hz). Moreover, the O–H stretching motion in H_2O molecules has further higher frequency ~3600 cm^{-1}, which corresponds to ~120 THz. A frequency of 10

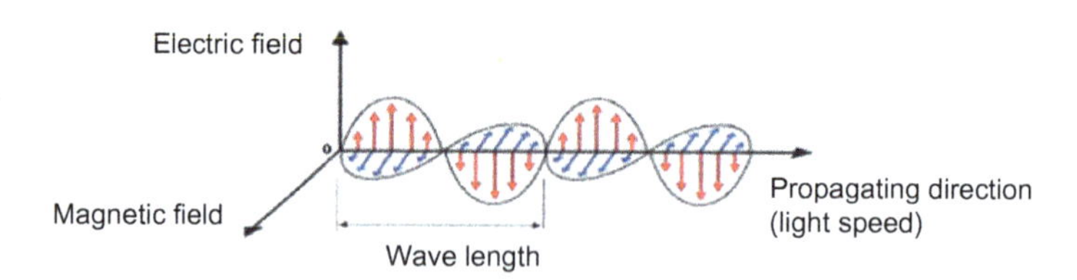

Fig. 3.3 Propagating electromagnetic waves in free space

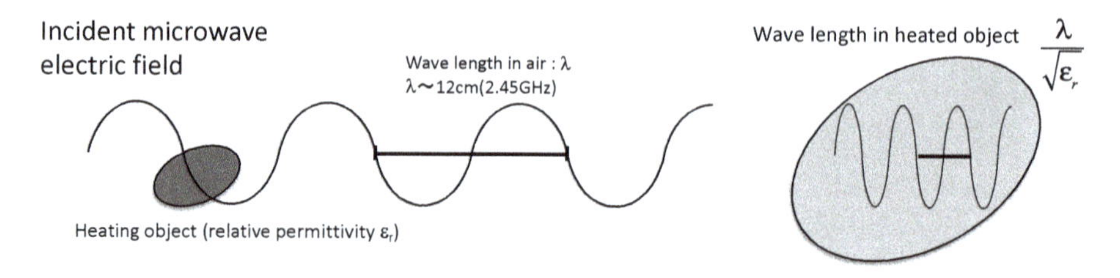

Fig. 3.4 Wavelength of the electromagnetic wave within a heated object

THz is much higher than that of microwaves, but heating of water up to 50 °C is far easily attained by microwave irradiation. How should we understand this fact? The likely answer is as follows: the distribution of the electron density (or their localization state) changes very quickly by the applied alternating electric field, following which the polarization of the molecules changes and leads to orientation polarization, then the rotational motion of the molecules is induced at the microwave frequency. Some "friction" among the neighboring molecules is expected as might occur upon rotational motion. These "frictions" or energy dissipation between molecules gives rise to the translational or higher frequency that we observe as "heat." Not only is this of interest for **Class I** materials but is also of importance on how we consider the interaction of microwaves with matter that leads to heating. This is the state-of-the-art in our understanding of "microwave heating." Let us now consider what the moving units are at the microwave frequency in **Class II** and **Class III** solids, and how they raise to the THz order of thermal vibrations. Thus far, the mechanisms are not well understood so that it is the purpose of the next few sections in this chapter to provide some description and discussion of these classes of materials.

3.3 Interaction Between Microwaves (Electromagnetic Waves) and Matter

As mentioned before that electromagnetic waves consist of oscillation of the electric (E)- and magnetic (H)-fields, it is to be noted, however, that the interaction of the E- and H-fields with matter is different from each other so that the manner with which the microwave energy is absorbed also differs for the E- and the H-fields. The energy absorbed by matter per unit time and volume, P [J m^{-3} s^{-1}], is expressed in terms of the magnitude of the E- and H-fields and their material properties ε'', μ'' and σ as expressed by Eq. (3.1):

$$\mathrm{P} = (\varepsilon''\omega|\boldsymbol{E}|^2 + \mu''\omega|\boldsymbol{H}|^2 + \sigma|\boldsymbol{E}|^2)/2 \tag{3.1}$$

where ε'' and μ'' are the imaginary parts of permittivity and magnetic permeability, respectively, and σ is the electrical conductivity. In the following sections, we intend to discuss the microwave heating mechanisms (or microwave energy absorption mechanism) from the view points of the interaction of the E- and H-fields with matter.

3.4 Heating Mechanism by the Microwaves' Electric (*E*-) Field

3.4.1 Molecules (Mainly Liquids)

Electronic cloud (electron density) is not homogeneously distributed within a water molecule but is localized in different ways between oxygen and hydrogen so that it forms an electric dipole (Fig. 3.5a, b). Multipole formation (such as a quadrapole) exists in some molecules; however, we restrict our discussion to only dipole because of the primary importance of dipoles. Some other molecules such as CO_2 acquire an electric dipole only when an electric field is imposed on the molecule. In this regard, because water molecule is inherently polarized, it is therefore called a permanent dipole.

When microwaves are imposed on water, rotational motions of the water molecules are raised owing to the alternating electric (*E*) field of the microwaves for orienting the dipole upon the imposed field direction (Fig. 3.5c). On the other hand, the overall motion of water (thermal motion) becomes more active as evidenced by an increase in temperature although the thermal motions occur randomly. When the temperature is relatively low, and the rotational motion of orienting molecules to the electric field is dominant (depending on the intensity of the *E*-field), it is gradually suppressed by the thermal agitation as temperature increases. Thus, the effect of microwave irradiation becomes weak for the cases of molecular liquids (**Class I** materials) as an increase of temperature. This phenomenon corresponds to the lower

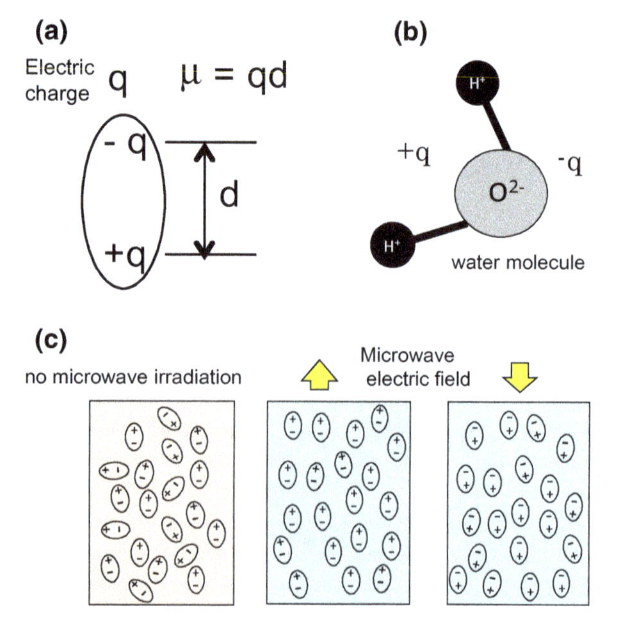

Fig. 3.5 Water molecules under microwave irradiation: **a** electric dipole, **b** polarization of the water molecule and **c** orientation of molecules under microwave application

dielectric loss (imaginary part of permittivity) values at higher temperatures [6]. This is an opposite tendency to the **Class II** materials as will be discussed later.

It is important to note that the induced motion of molecules by an electromagnetic wave depends on its frequency. For example, irradiation with ultraviolet waves (usually not called electromagnetic waves if they were above a terahertz frequency; they would be regarded as light) interacts with the motion of electrons existing in the molecular orbitals that are responsible for the binding of atoms. Irradiation with ultraviolet rays gives rise to distortion of the electronic clouds (distributions), thereby inducing electronic polarization. Because the frequency of ultraviolet radiation is so high, only electrons (have lighter weight) can respond quickly to the field variation.

On the other hand, infrared waves have a lower frequency (longer wavelength), and so, it is able to induce such atomic (ionic) motion as the stretching of atomic bindings or translational motions of molecules. These are referred to as atomic or ionic polarization, and the molecular motions are directly related to the heat or are comparable to the thermal vibration of molecules. Nevertheless, as mentioned above, microwaves can only raise the rotational motion of the molecules having the same frequency order. The molecules are oriented toward the applied electric field of the microwaves by rotating the dipoles, referred to as orientation polarization.

The frequency spectra of polarizations are illustrated in Fig. 3.6, which are expressed in terms of real (upper) and imaginary (lower) parts of the permittivity. This spectrum is not limited to **Class I** materials but also applies to **Class II** materials. The dielectric response due to electrons, atomic (ionic) and orientation polarization exists from the higher frequency in this order.

As depicted in Fig. 3.6, we recognize that the spectra of electronic and atomic polarizations differ from orientation polarization in that the peak widths of the imaginary parts of the former are very sharp, but the peak of the latter is much broader. Moreover, the real part of the orientation polarization (ε') gradually decreases as the frequency increases, while the imaginary parts of other polarizations have a different form.

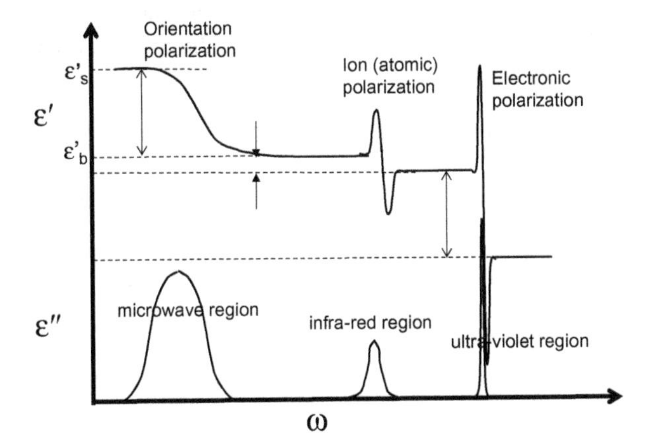

Fig. 3.6 Polarization (permittivity) spectrum and electromagnetic wave energy absorption

The differences of these behaviors find their origin from the difference in the mobility or states of the response to the external electric field. It can be interpreted in terms of the one-dimensional oscillator model, which consists of a spring and a weight, as schematically illustrated in Fig. 3.7 (vibration of an electric charge is modeled with one-dimensional oscillator, which also expresses a motion of the electric dipole). Oscillation (vibration) is excited by an external force; friction with the floor is taken into consideration. The motion of the electric dipole is regarded as the vibration of a weight of mass m bound to a spring (spring constant k). The manner of vibration is determined by the relative magnitudes among the forces of the inertia, the friction force, the restoring forces. If there are no external force nor friction, the system has the Eigen frequency (ω_0) determined by k and m ($\omega_0^2 = k/m$). When the frequency of the external electric field matches with ω_0, resonance occurs. Even if a friction force exists, the resonance type of vibration occurs under the conditions that the inertia and the restoring forces balance each other and with relatively lower friction force. The spectrum of the imaginary part of the permittivity becomes a Lorentzian-type (resonance type), and the peak has a sharp form. On the other hand, under the conditions that the inertia force can be ignored with respect to the restoring force, and that the friction force is relatively high, its functional form becomes a Debye-type (relaxation-type), and the peak becomes broader. The polarization relaxation occurring at a higher frequency (range of light wavelength; electronic and atomic polarization) becomes of the resonance type, and the orientation polarization becomes of the relaxation-type.

At this point, it is relevant to inquire as to whether there are any differences between microwave heating and heating at other frequency ranges in Fig. 3.6. With the infrared waves, it is possible to excite the translational motion of molecules directly and heat the matter, but just only on surface. In microwave heating, however, the range of the heating area (penetration depth) is larger. That is, because the frequency of infrared waves (order of 10^{12}–10^{14} Hz) is about an order of 10^4 higher than that of microwaves (assuming a frequency of 10^9 Hz), the penetration depth becomes 1/100. For instance, if a 3-cm thick steak was heated into the internal regions by the microwaves, infrared waves can penetrate only to a ca. 300 μm thickness on the surface (in reality, the temperature of the internal area can also be raised by thermal

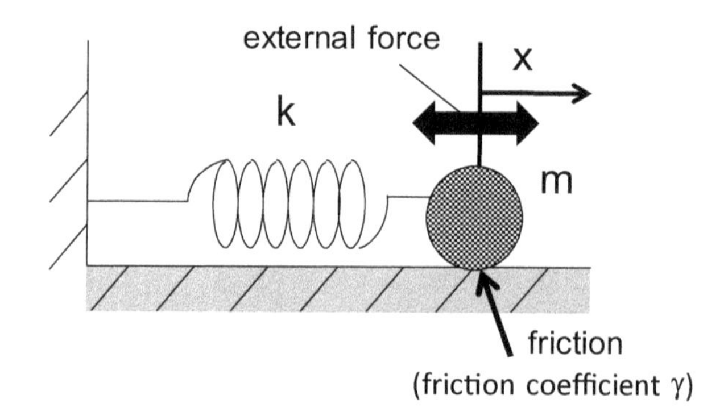

Fig. 3.7 One dimensional oscillator model of an electric dipole

conduction). On the other hand, what about irradiation of matter with visible light or ultraviolet radiation (having much higher frequency)? In this case, it is not possible to raise the temperature directly because their quantum energy is much higher than that of microwave or infrared radiation, and thus, they are more likely to influence only the atomic bonding of the molecules.

3.4.2 Inorganic Solids

Polarization of molecules (**Class I**) exists and/or occurs; however, it is not limited to molecules, as **Class II** materials such as glasses and ceramics (non-conducting materials having ionic and covalent bonding) also exhibit polarization. Electronic and atomic (ionic) polarization can be discussed in the similar way as **Class I** materials. It is easy to imagine the individual motions of molecules in **Class I** materials, such as rotation, translational and stretching. But now question is what moves in the inorganic solids when exposed to the electric field at the frequency of the microwave? In other words, what are moving and causing orientation polarization (heat generation)? It is not necessarily easy to have conceptual images to specify the mobile units. According to the vibration model, motion of mobile units having at least a larger mass than electrons and atoms (ions) must be excited. And both restoring and friction forces are also expected to be exerted in their motions. Some group of atoms or atomic clusters in inorganic solids are to be taken into consideration; however, it is not easy to specify nor to identify as stated above. As one of the clues to imagine the mobile units, vacancies in ionic crystals are illustrated schematically in Fig. 3.8. The ionic crystal lattice contains ion vacancies, the concentration of which increases as the temperature increases according to the thermal equilibrium. The ions are allocated in the crystals so that electrical neutrality is fulfilled; however, if ion vacancies are introduced into the crystal, the neutrality conditions become unsatisfied locally. This area could be regarded as an electric dipole. If some ions jump into the other sites, the dipole orientation changes, as can be seen in Fig. 3.8a, b. This could be regarded as the rotational motion of the electric dipole. A vacancy does not necessarily exist in a single state as a vacancy cluster is also possible to occur. The states and the scales of the real electric dipole could be more complicated and have more variations.

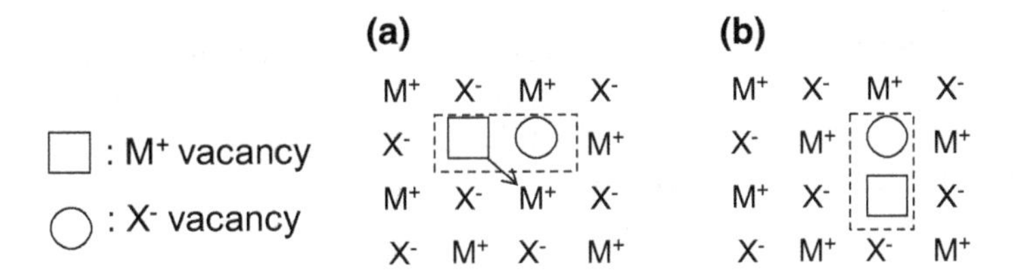

Fig. 3.8 Formation of electric dipole in ionic crystal having vacancies. M^+ vacancy (**a**) jumping to the oblique site, corresponding the rotation of electric dipole (**b**)

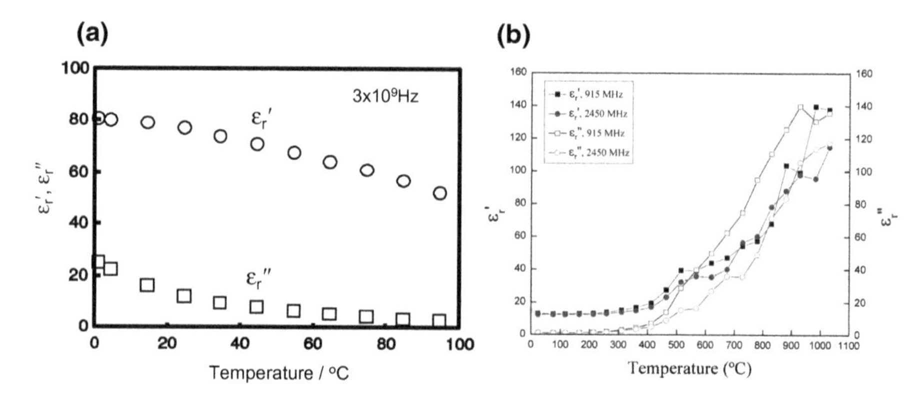

Fig. 3.9 Permittivity dependence on temperature: **a** water [6] and **b** Fe_3O_4 permittivity [7]

The temperature dependence of the dielectric relaxation by orientation polarization of **Class I** materials was discussed before as follows: Molecules of **Class I** materials rotate rather freely, thermally excited rotational motion become more active at higher temperatures. Irradiation with microwaves excites the orientation but is hindered by the thermal agitation. Therefore, the polarization or the permittivity of **Class I** materials (e.g., water) decreases as the temperature increases as shown in Fig. 3.9a [6]. On the other hand, the temperature dependence of **Class II** materials is opposite. A temperature increase causes an increase in vacancy concentration and in their mobility (jumping frequency) so that the polarization or the permittivity increases at higher temperatures. As an example, the temperature dependence of the permittivity of Fe_3O_4 is plotted in Fig. 3.9b [7].

Water has a boiling point of 100 °C so that the temperature range for consideration is less than 100°; however, the ceramics or glasses have a larger range up to more than 1000 °C. Usually, above 500 °C, a large increase is observed as exemplified by the case of Fe_3O_4.

3.5 Mechanism of Heating by the Microwaves' Magnetic (*H*-) Field

This section will describe and discuss electric conductive materials such as metals and ferro/ferrimagnetic materials. In particular, *H*-field heating effects will be discussed in two cases of materials separately. In this regard, iron satisfies both requirements and is thus heated well by microwaves.

3.5.1 Electric Conductor

Let us first consider an experiment whereby a metal is placed between two electrodes, following which an electric voltage is imposed between the electrodes that are not in contact with the metal (Fig. 3.10). When DC voltage is imposed, nothing happens to the metal because the electrons in the metal keep their homogeneous density distribution. Similarly, if an alternating electric field was applied, the situation would be the same. Because electrons in a metal move very rapidly, they will often collide with each other, the frequency of which may be around 10^{13} s^{-1}; therefore, the alternating electric field cannot influence the electrons below the electron collision frequency. However, if we were to replace the metal with a dielectric material, then polarization would occur and something (dielectric relaxation, etc.) would then happen. Therefore, it is expected that if an alternating electric field at the microwave frequency was applied to the metal, nothing would seem to change. Note that this is a "thought experiment"; in reality, it is not possible to impose only an alternating electric field at high frequency because it would create an alternating magnetic field in the metal simultaneously, which would generate induction current.

On the other hand, if an alternating magnetic field was applied to the metal, it would be very effective to generate an induction current because it would generate an electric field gradient in the metal by which electrons are driven to move. This is the principle of induction heating and is applied to IH (induction heating) cooking heater (the frequency is of the order of kHz, having the skin depth of several millimeter and comparable to the pan thickness). The penetration depth of an alternating magnetic field decreases with an increase of frequency so that 2.45 GHz microwave induction heating of metal is limited within a depth of 1 μm or less. Therefore, it is difficult to raise the temperature of bulk metal because if only the surface thin layer were heated, the heat would dissipate into the internal area due to high thermal conductivity, and the temperature of the whole metal bulk would not be raised. Only metal powders or thin films are effectively heated by microwaves.

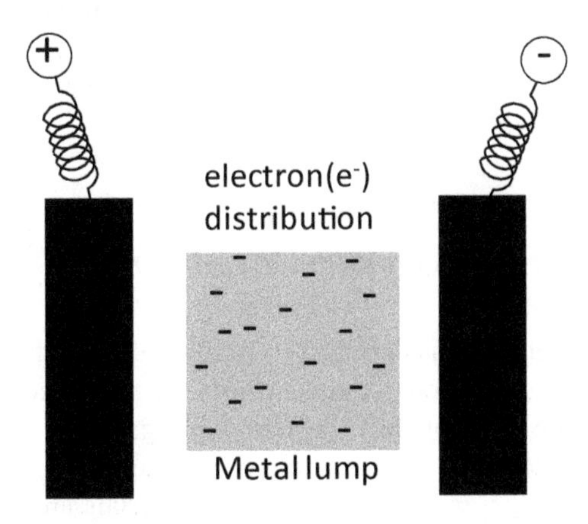

Fig. 3.10 Metal lump placed between two electrodes

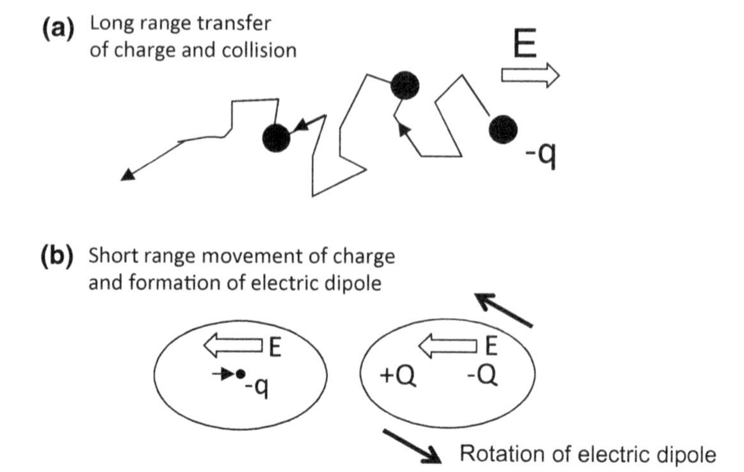

Fig. 3.11 Conceptual figures of **a** conduction (ohmic) loss, and **b** dielectric loss by rotation of electric dipoles

As mentioned above, the experiment of applying either a single electric or magnetic field at high frequency is not possible by the setup shown in Fig. 3.10. However, it is possible to accomplish it in separated *E*- and *H*-field applications by using a microwave cavity (In the RF frequency case, the cavity size becomes very large and is therefore not usually used.).

It is worth noting, however, that metal heating occurs more effectively when using a *H*-field than when using an *E*-field. The induction current in metals is the movement of free electrons which are driven by the electric field gradient created by imposition of the alternating magnetic field. The distance of the electron movement is not within the localized space around the atom (positive ion), but electrons can move long distance such that heat is generated by their interaction with the thermal vibration or with impurity atoms, referred to as electric resistance and Joule heating (conduction or ohmic loss). This is different from the electron motion localized in the neighborhood of atoms in **Class I** or **Class II** inorganic materials. The difference in the states of electron behavior is schematically illustrated in Fig. 3.11 ($q = e$, where e is the electron charge).

3.5.2 Ferromagnetic Materials

Before considering the microwave heating of magnetic materials, let us recall the classification of magnetization of substances as the states of electron spin alignment are schematically illustrated in Fig. 3.12. Paramagnetic materials have randomly oriented spins (Fig. 3.12c) but are magnetized (aligned) to the direction of the imposed magnetic field. Diamagnetic materials are magnetized in the opposite direction to

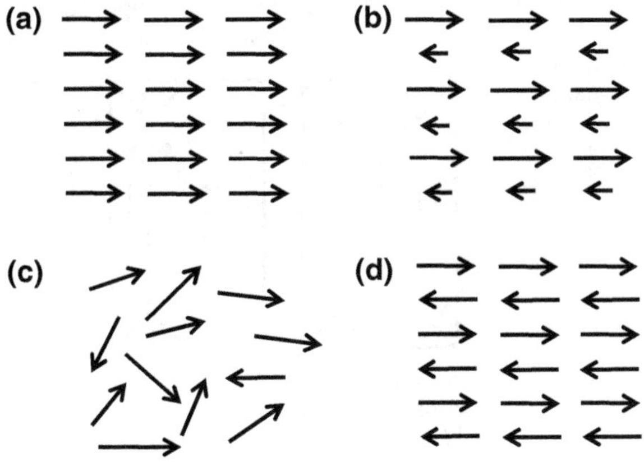

Fig. 3.12 Schematic images of spin alignment in various magnetic states. **a** Ferromagnetism, **b** ferrimagnetism, **c** paramagnetism, **d** antiferromagnetism

the field. For example, oxygen and nitrogen gases are paramagnetic and diamagnetic molecules, respectively. Solid metals of gold and copper are paramagnetic and diamagnetic systems, respectively. However, the degree of their magnetization is very small, and thus, we do not recognize that "they stick to a magnet," unless a strong magnetic field is specifically imposed such as above 10 Teslas (10 T). They are said to have weak magnetization. On the other hand, iron sticks to a magnet and becomes a magnet when it is magnetized. Solid iron, nickel, cobalt, some alloys and some rare-earth metals are classified as ferromagnetic materials, and their spins are inherently aligned together (Fig. 3.12a). Their degree of magnetization (magnetic susceptibility or magnetic permeability) is much larger than that of paramagnetic and diamagnetic materials. They have a strong magnetization. The nature of the ferromagnetism originates from the unpaired outermost electron spins. They are spontaneously aligned together because of the interaction known as exchange forces, but they are divided into the magnetic domains where spins are aligned in the same direction. The spin (domain) states of the strong magnets are schematically illustrated in Fig. 3.13.

There are two types of strong magnetization states. One is a ferromagnetic state (Fig. 3.12a), and the other is a ferrimagnetic state. Ferrite (iron oxide, for example) is a typical example of the latter, as the spin state is shown in Fig. 3.12b. Although the outmost spin alignment of iron and oxygen is opposite, the magnitude of the spin momentum of iron is larger than that of oxygen. Therefore, the material eventually exhibits spontaneous magnetization as ferromagnets.

The last class consists of antiferromagnetic materials (Fig. 3.12d). In this case, the magnitudes of the opposing spins are the same and cancel each other out; hence, they also exhibit weak magnetization. Chromium (Cr) metal is of this type.

As mentioned above, ferromagnetic and ferrimagnetic materials possess magnetic domains that are separated with domain boundaries (Fig. 3.13). When a magnetic field is imposed, the volume of the domain where spins are aligned to the field direction increases and the domains come to possess the same direction of spin alignments. This is the magnetization process of ferromagnets (ferrimagnets; later on, we shall refer to these simply as ferromagnets). In this manner of magnetization, if an alternating

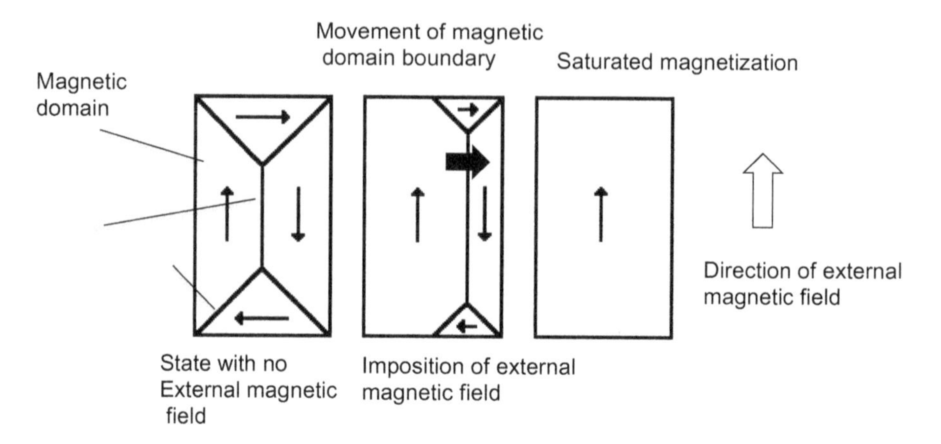

Fig. 3.13 Magnetic domain structure and magnetization

magnetic field is imposed, the domain boundary has to move and follow the variation of the magnetic field. The resistance of the boundary movement becomes one of the causes of high-frequency magnetic loss. However, at the microwave frequency, the domain boundary cannot generally keep up with the field change. In reality, other phenomena also occur, which are not a simple process but experience through many complicated mechanisms.

Ferromagnetic resonance (FMR) is one of the phenomena to be taken into consideration as one of the microwave magnetic heating (loss) mechanisms. When a static magnetic field is imposed to the ferromagnets, precession motion of electron spins occurs along the field direction as shown in Fig. 3.14. The frequency of the precession motion is proportional to the magnitude of the external magnetic field, and resonance (FMR) occurs when it matches with that of the irradiated microwave frequency. FMR is the equivalence of the electron spin resonance (ESR) in ferromagnetic materials, and the one that occurs in paramagnetic materials is particularly called electron paramagnetic resonance (EPR). The intensity of energy absorption

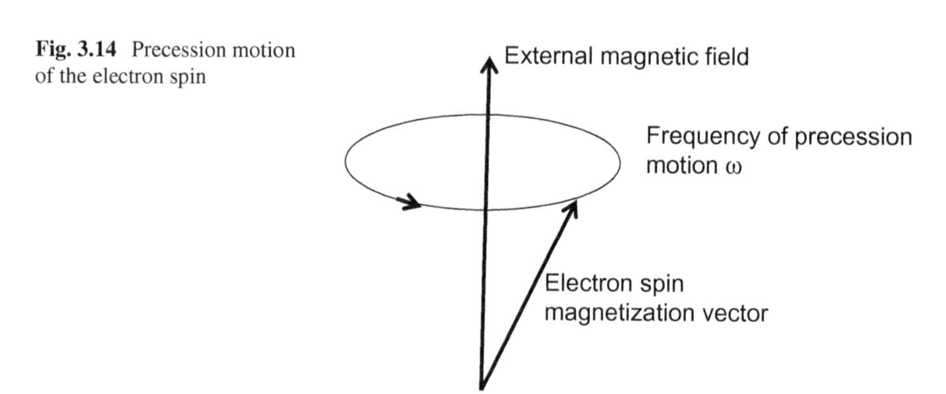

Fig. 3.14 Precession motion of the electron spin

of FMR is much larger than the general ESR (EPR as well) and is considered to be related with microwave magnetic loss of the ferromagnetic materials.

Experimental results of microwave FMR heating by imposition of an external static field is shown in Fig. 3.15 [8]. In this experiment, a sample of compressed powder Fe_3O_4 ($\phi = 7$ mm, $t = 0.5$ mm) was placed at the *H*-field maximum position in a cavity, after which it was irradiated with microwaves that led to an initial increase in temperature up to 420 °C as can be seen in the time chart displayed in Fig. 3.15a; subsequently after 250 s, a static magnetic field was imposed perpendicular to the oscillation direction of the microwave *H*-field, increasing its magnitude up to 0.5 T gradually, which caused a temperature increase of about 45° (having a peak at $B(\mu_0 H) = 0.17$ T) and then decreased even though the *H*-field increased up to 0.5 T at 650 s after starting the experiment. After the imposed field reached maximum, the *H*-field intensity was decreased down to zero at a constant rate although we observed the temperature peak at the same *H*-field (Fig. 3.15b) again. The temperature increase in both ascending and descending *H*-field must be caused by FMR, indicating that the FMR could be one of the key magnetic loss mechanisms in microwave processing, considering certain aspects described below.

FMR occurs in ferromagnetic materials even without applying a static magnetic field. There are many kinds of internal magnetic fields in ferromagnets, such as a crystalline magnetic anisotropy field and a demagnetization field, among others. The spin precession motion occurs in a wide frequency range from several MHz to hundreds of GHz, interacting with the electromagnetic wave and giving rise to FMR, depending on the internal magnetic fields in different materials. This is a

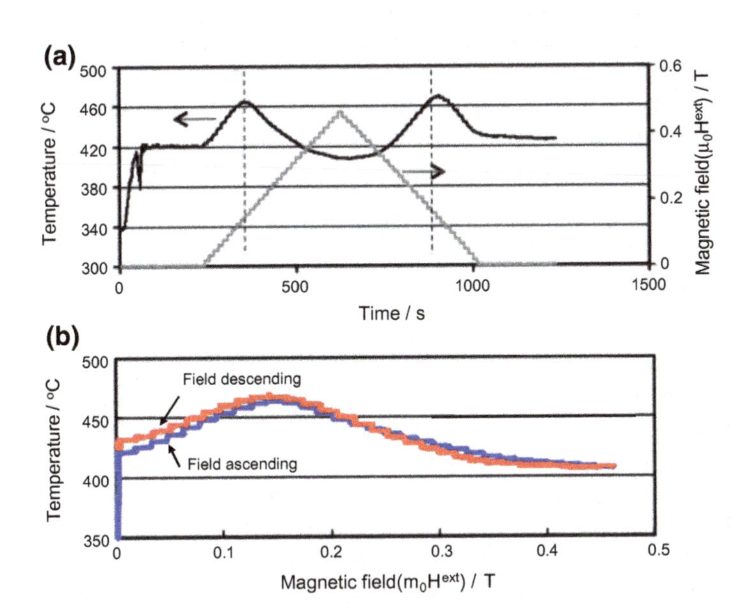

Fig. 3.15 Magnetic loss due to FMR [8]. **a** Temporal variation of temperature and the magnetic field intensity. **b** Dependence of temperature on imposed magnetic field

well-known phenomenon of causing a loss of the high-frequency ferrite, which is termed as "natural resonance." Therefore, if a high-frequency electromagnetic wave irradiated ferromagnets, natural resonance takes place, and more or less absorption of microwave energy occurs without the imposition of an external magnetic field.

3.5.3 Distinction Between Induction Current (Ohmic) Loss and Magnetic Loss

Some confusion exists in understanding the microwave magnetic heating mechanism (magnetic loss). If we apply an alternating magnetic field to a composite material consisting of a weak magnetic metal particles (for example, paramagnetic Al), which are dispersed in an Al_2O_3 ceramic matrix, heating occurs because of the induction current generated in the metal particles (Al_2O_3 at room temperature is not lossy). If its magnetic permeability is measured at high frequency, its imaginary part (μ'') would exhibit a non-zero value owing to a loss caused by an induction current. It looks as if Al is a ferromagnet. The apparent μ'' of paramagnetic Al is caused by the alternating H-field. The heating mechanism is nothing but that of an induction current loss (third term in Eq. 3.1). There are many researches who consider this phenomenon as a magnetic loss mechanism.

We are at a standpoint that the true magnetic loss corresponding to the second term in Eq. 3.1 is some magnetic mechanism in which the electron spin is involved, that is, once the microwave energy is converted into changes of the electron spins, it is then transferred to the lattice as FMR or as the domain wall resonance, among others. Therefore, this magnetic loss can be distinguished from the induction current loss. The latter is just a Joule heating effect. In reality, their separation and evaluation of the contributing ratio are not simple experimentally.

3.6 Converting Mechanisms of Microwave Energy into Heat

So far, we have discussed the microwave heating energy loss mechanisms in different matter. The effects of a microwave electric field are to excite the rotational motion of electric dipoles in molecules that possess a dipole moment (**Class I**) or some atomic clusters that possess a dipole moment (**Class II**). The effects of a microwave magnetic field are to generate an induction current in electrical conducting materials (**Class III**, metals and conductive ceramics among others) or to excite magnetic spin-related phenomena, the energy of which is then transferred to the lattice.

As discussed in Sect. 3.2, the heat or temperature increase observed in matter is either by translational motion of the molecules or vibration of the crystal lattice or atoms (ions, in the glassy states), the frequency of which is in the order of THz (i.e.,

10–100 THz), and much higher than the frequency of microwaves by three to four orders of magnitude. How are these higher frequency vibrations excited by the lower energy stimulation of microwaves? Let us consider them as follows:

- The microwave electric field excites rotational motion of electric dipoles of molecules to be oriented in the direction of the field. However, there are some mutual interactions among the molecules. If one single molecule rotates to the energetically favorable dipole direction with respect to the electric field, it has to overcome the interaction force, which could play a role of "friction" to the motion. Although the nature of this interaction is of interest, it is not necessarily easy to be identified nor to be investigated.

It is naturally anticipated that this "friction force" influences not only the rotational motion itself, but it causes some extra motion of the molecules or some vibration in the molecules. This extra motion could be the translational motion or a higher frequency motion observed as heat, the status of which is schematically illustrated in Fig. 3.16. A molecular dynamic study of water molecules under irradiation by microwaves [9] has indicated the existence of some kinetic spectrum at higher frequency, which might be related with the "friction" among the molecules.

- When the induction current is generated in metals (**Class III**), electron motions are enhanced. Then, the electric current raises the lattice vibration (in the quantum mechanical picture, the electron wave interacts with the wave of the lattice vibration (phonon)). When the lattice vibration is enhanced, temperature increase or heat generation occurs. As stated above, the lattice vibration frequency is much higher than that of microwaves. Here, we have to remember the electric current in metals generating heat is not only caused by RF power or by microwaves but by application of a direct current. Therefore, as noted before, microwave induction current loss heating is just a Joule heating, and the heating mechanism of which is

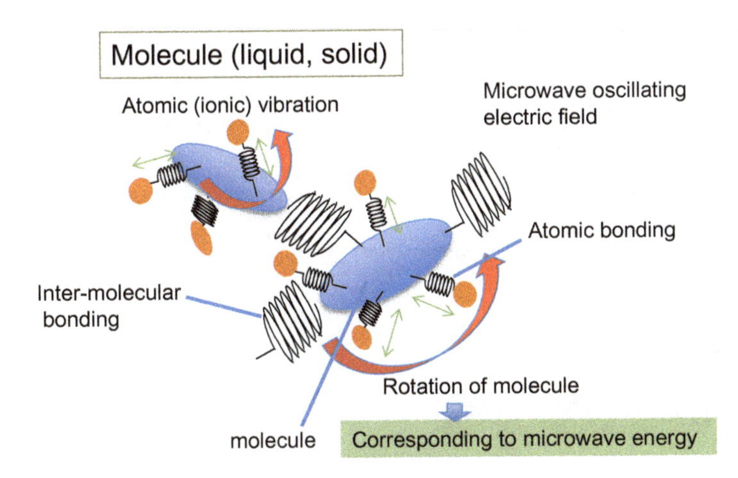

Fig. 3.16 Conceptual image of vibration of molecule related with heating and with absorbing microwave

the same as the general electrical heater with a nichrome wire operated at a commercial alternating frequency although the area of the electric current generated at the microwave frequency is limited just on the surface area.

In **Class II** materials, it is not possible to excite the individual atomic (ionic) motion by microwave irradiation because the frequency of the atomic (ionic) motion is much higher (in the infrared region). However, microwave irradiation may excite (the rotation) the motion of some "units or clusters" of atoms (ions), which have electric dipoles (having larger scale or weight). Upon excitation of these larger dipole motions, it is expected that the motion of the unit will also induce the "friction" or interaction with the surroundings, which is also responsible for generating the high-frequency vibration of the lattice (phonon). Although it is not necessarily easy to determine the moving "unit" experimentally, it would be helpful to present a conceptional image of the mechanism by an analogous picture shown in Fig. 3.17 (a similar picture was also shown in Ref. [10]). In Fig. 3.17, the leaves are connected to the small branches of a large tree with rigid springs. When a strong wind shakes the tree, the main tree trunk and the branches experience shaking having a rather long period, while vibrational motions of the leaves are also excited. The period of the leaves' vibration must be much smaller than that of the branches because the binding with the rigid springs is assumed. The shaking frequency corresponds to that of microwaves and the frequency of the leaves to the lattice vibration (heat).

In the case of magnetic loss, the microwave energy is converted into electron spin motions. Because the electron spin has a mutual interaction with the lattice and the

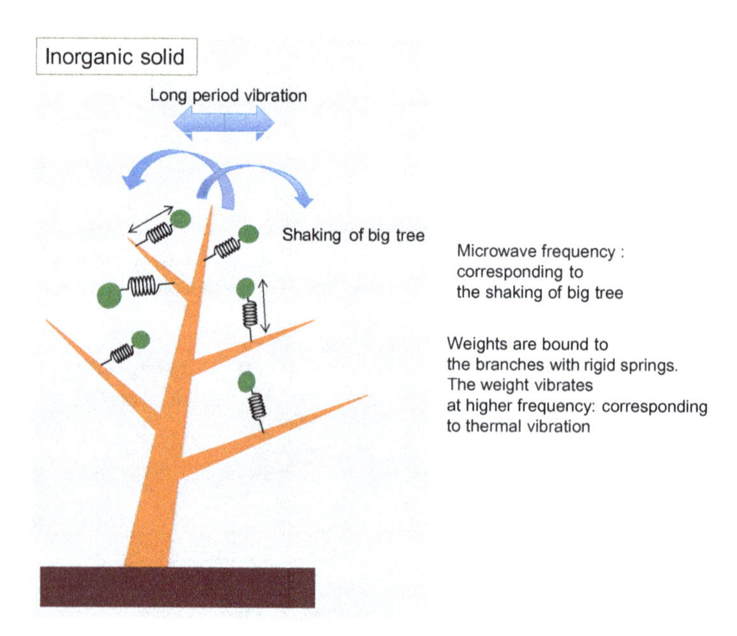

Fig. 3.17 Conceptual image of the relationship between the motion excited by microwaves and the heat (temperature increase) generated in the object

lattice vibration, it is very likely that the energy of the spin motion is eventually transferred to the lattice. In ESR and NMR schools, they measure and discuss the time periods of spin–spin relaxation and the spin–lattice relaxation. The latter is expected as a heating phenomenon in the magnetic resonance and is related to the FMR heating experiments shown in Fig. 3.15. On the other hand, it is known that a magneto-elastic interaction is possible to excite the ultrasonic (elastic) waves at a microwave frequency through magnetostriction. This is a direct interaction between the electron spin (magnetic) energy and the lattice motions. In this case, however, special conditions that accord with spin wave frequency and the wavenumber (inverse of wavelength) have to be fulfilled in order to excite the elastic wave effectively. In order to overcome the difference in the phase velocity between the spin wave and the elastic wave, use of thin films and their standing wave conditions are utilized [11]. The ultrasonic wave of microwave frequency is still lower than that of lattice vibrations. The ultrasonic energy is eventually also converted into heat. Some different processes of converting the ultrasonic energy to dissipate into heat (thermal energy) must be happening.

References

1. Metaxus AC, Meredith RJ (1983) Industrial microwave heating. In: IET power and energy series 4, 1st edn. The Institute of Engineering and Technology, London
2. National Research Council (1994) Microwave processing of materials. National Academy Press, Washington D.C.
3. Whittakar AG, Mingos DM (1995) J Chem Soc (Dalton Trans) 12:2073–2095
4. Clark DE, Folz DC, West JK (2000) J Mater Sci Eng A287:153–158
5. (a) Yoshikawa N (2014) Fundamentals in microwave processing of materials. Corona-sha, Tokyo, Japan. ISBN-10: 433904637X (in Japanese); (b) Yoshikawa N (2010) J Microwave Power Electromagn Energy 44:4–13
6. von Hippel AR (Ed) (1954) Dielectric materials and applications, Chap. V. MIT Press, Cambridge, Mass, pp 291–429
7. Peng Z, Hwang JY, Mouris J, Hutcheon T, Huang X (2010) ISIJ Int 50:1590–1596
8. Yoshikawa N, Kato T (2010) J Phys D Appl Phys 43(425403):1–5
9. Tanaka M, Sato M (2007) J Chem Phys 126(034509):1–9
10. Booske JH, Cooper RF, Dobson I (1992) J Mater Res 7:495–501
11. Kittel C (1958) Phys Rev 110:1295–1297

Chapter 4
Microwave Flow Chemistry

Joshua P. Barham, Emiko Koyama, Yasuo Norikane and Takeo Yoshimura

Abstract This chapter presents examples of and advocates for the adoption of tunable solid-state (semiconductor) oscillator single-mode microwave flow reactors toward laboratory and larger-scale synthetic chemistry applications. Tunable solid-state oscillator single-mode microwave flow reactors are more versatile heaters that impart both better process control and energy efficiency than conventional magnetron oscillator flow reactors when operated in single-mode or multimode.

4.1 Introduction

4.1.1 Microwave Heating Devices

The domestic microwave oven uses microwaves for heating across the globe. Microwaves generally refer to radio waves with frequencies of 300 MHz–300 GHz (wavelengths of 1 m–100 μm) [1–3]. In practice, the available frequency for microwave heating is defined as the Industry Science Medical (ISM) band by the International Telecommunication Union (ITU) [4]. The global microwave ISM bands are 2.45, 5.8 and 24 GHz (there are other, locally permitted microwave bands in certain global areas, such as 0.915 GHz, allowed in North and South America). The law dictates that microwaves can only be employed for heating devices between 2.400 and 2.500 GHz [1] and so the 2.45 GHz (±0.10 GHz) band is widely used. However, the materials ('loads') for which heating by microwaves is desirable have various dielectric constants and their microwave absorption characteristics differ for each frequency [5, 6].

J. P. Barham (✉) · E. Koyama · Y. Norikane
National Institute of Advanced Industrial Science and Technology (AIST), Central 5, 1-1-1 Higashi, Tsukuba, Ibaraki 305-8565, Japan
e-mail: j.barham@saidagroup.jp; Joshua-Philip.Barham@chemie.uni-regensburg.de

T. Yoshimura (✉)
SAIDA FDS Inc., 143-10, Isshiki, Yaizu, Shizuoka 425-0054, Japan
e-mail: t.yoshimura@saidagroup.jp

S. Horikoshi and N. Serpone (eds.), *RF Power Semiconductor Generator Application in Heating and Energy Utilization*, https://doi.org/10.1007/978-981-15-3548-2_4

The penetration depth (D_p) of microwaves in materials is defined by Eq. 4.1 [3]. This is the depth at which the electric power becomes half that at the surface of a given load, which differs depending on the imaginary and complex components of the permittivity, ε' (the dielectric constant) and ε'' (the dielectric loss). Naturally, Eq. 4.1 infers a limitation on the size (or scale) of heating [7]; for example, the microwaves' (2.45 GHz) penetration depth in water at 25 °C is only 1.4 cm [3]. Therefore, precise microwave heating requires consideration of scale even for the same material. A microwave heating device comprises of three components: (i) an oscillator, (ii) an applicator or 'cavity' and (iii) a reaction vessel. The appropriate combination of these three components is key to the performance of the heating system.

$$D_p = \frac{\lambda}{2\pi}\frac{\sqrt{\varepsilon''}}{\varepsilon'} \tag{4.1}$$

The oscillator is a device that generates microwaves. The most common type of oscillator is a magnetron; an inexpensive electron tube that emits a range of frequencies distributed on each side of the center frequency. Some of these frequencies may couple to the resonant frequencies of the load, thereby effecting heating via dielectric loss. A less common type of oscillator is the solid-state (semiconductor) device, which has recently emerged as a microwave generator in a variety of industries, particularly in the telecommunication industry. This device consists of an oscillator and an amplifier. The solid-state device is expensive compared to a magnetron; however, it emits a specific frequency which is controllable ('tunable') and it benefits from a considerably longer lifetime compared to the magnetron [8–13].

The cavity's role is to confine electromagnetic waves. There are two types. The first type is a 'traveling wave cavity', where microwaves make a single pass through a load, are reflected off the cavity wall and make a return pass through the load. The traveling wave cavity therefore has a low power density; The second type is a 'resonator cavity', where a wall or plunger is set to reflect the microwaves, allowing propagation of the standing waves within the cavity to create a local maximum [14]. The resonator cavity therefore has a high power density. Both these cavity types can be operated in 'single-mode,' which means that a definable microwave waveform is generated in the cavity. Alternatively, they can be operated in 'multimode,' which means that a broad range of microwave frequencies and waveforms are allowed to distribute randomly in the cavity [15, 16]. Some waveforms will couple (interfere constructively or destructively) with available resonant frequencies of the load to be heated, leading to 'hot and cold spots.' The domestic microwave oven is a resonator cavity operating in multimode. The multiple random electric field modes in the resonator cause uneven heating, which is traditionally mitigated by rotating the load on a turntable. Recent microwave ovens are often equipped with 'a mode stirrer'—a device that rotates the electromagnetic field distribution [15–17].

With regard to the reaction vessel, its size and the material that it is made of have a large influence on microwave heating. The 'dielectric loss tangent,' tan δ (Eq. 4.2), indicates the ease of microwave absorption [2, 15]. Small tan δ values (e.g., that of

Quartz) indicate that vessels have poor absorption, which renders them challenging to heat under microwave irradiation—nonetheless, it follows that heating of the vessel's contents may occur selectively [18, 19]. On the other hand, large tan δ values (e.g., that of SiC) indicate that vessels have high absorption, thus rendering their facile heating under microwave irradiation—it follows then that heating of the vessel's contents may occur by conductive heating resulting from the microwave heating of the vessel walls [19–22]. An interesting question is then: When the cavity and the reaction vessel components of the microwave device are identical, what impact does the choice of oscillators have on microwave heating?

$$\tan \delta = \varepsilon''/\varepsilon' \tag{4.2}$$

Herein, we propose that the greatest power of the oscillator lies in its ability to vary the frequency. Solid-state (semiconductor) oscillators irradiate at a single frequency, are energy efficient, and their irradiation frequency can be changed. On the other hand, the magnetron has a wide oscillation frequency bandwidth and cannot emit an isolated single frequency. Generally, the oscillation frequencies of a magnetron cannot be changed. Since a solid-state oscillator (used in conjunction with a single-mode cavity) can be tuned to detect the frequency at which microwave absorption is highest, the heating efficiency is generally very high with respect to the magnetron. Under certain conditions, the heating efficiency of a 100 W power solid-state device (used in conjunction with a single-mode cavity) can be several times higher than that of a 700–1500 W domestic microwave oven. Despite the excellent characteristics of solid-state devices, magnetron devices are significantly cheaper to manufacture and are reasonably priced for consumers. This explains the prominence of the magnetron in a wide range of fields such as drying and heating. Nevertheless, certain heating applications benefit more from the use of solid-state devices for microwave generation, such as in chemical synthesis, and in particular in 'flow chemistry' (Sect. 4.1.3).

4.1.2 Microwave Heating in Chemical Synthesis

Microwaves have been used to carry out chemical reactions since 1986; it is known that microwave heating can shorten reaction times and improve yields compared to conventional heating [23]. However, the use of microwave heating in synthetic chemical industries has been hindered because (a) such industries require large-scale applicability, and (b) even heating can be critical in ensuring process control, reliability and product quality.

Regarding large-scale applicability, there are several issues in the scale-up of batch microwave processes. Owing to the previously mentioned penetration depth issue (Eq. 4.1), large-scale processes are difficult to carry out in batch reactors because only the solution close to the vessel walls may undergo microwave heating. The previously mentioned 1.4 cm penetration depth of water (at 2.45 GHz and 25 °C) highlights the practical challenges in constructing a batch microwave apparatus of several hundred liters. Moreover, the yield of a given chemical reaction varies unpredictably as the reaction scale is changed.

Control, reliability and distribution of heat are directly related to the electromagnetic field distribution and strength, which vary depending on the type of apparatus employed. Commercially available instruments using conventional multimode operation and magnetron oscillators have poorly characterized electric field distributions, which compromise the reproducibility of chemical reactions. Single-mode cavities with more defined electromagnetic distributions are desirable, but the ISM band (2.40–2.50 GHz) restricts the available microwave frequency range and imparts a big restriction on the cavity size (the wavelength of 2.45 GHz is 12.2 cm), such that scale-up of batch processes necessitates multimode cavities and magnetron oscillators. However, in addition to the poorly characterized electric field distribution, the conversion of electrical energy into microwave energy by magnetrons is relatively inefficient (ca. 70% or less), which represents a significant concern to process chemistry departments [24–26].

An alternative method for scale-up is 'flow chemistry,' which has gained prominence for chemical synthesis in recent years. Herein, 'flow chemistry' does not refer to the high-volume, thick-pipe production of petrochemicals and bulk chemicals—rather, it refers to a method for pharmaceutical, fine chemical or material synthesis requiring precise heating control and using pipe diameters of several mm to several cm. Accordingly, it is now useful to describe some principles and benefits of flow chemistry.

4.1.3 Flow Chemistry: Principles and Benefits

'Flow Chemistry' is the processing of a chemical reaction in a continuously flowing stream, using a tubular reactor instead of a batch-type flask reactor. Regarded globally as a promising technology, flow chemistry is already established in multiple synthetic chemical industries such as pharmaceuticals, functional materials, petrochemicals and fine chemicals [27–32]. The benefits of operating chemistry in a continuous flow instead of a batch reactor are clearly understood.

Cost: Because the scale capability and productivity of a flow reactor depends on the flow rate, which is not necessarily linked to reactor size, flow reactors are generally smaller than batch reactors for the same given productivity [33]. Therefore, facilities can be more compact, require less capital investment and have lower maintenance costs. In addition, should a deviation occur which compromises the quality of a flow process, the reaction volume to be discarded is smaller and therefore less costly than that of a batch process [29].

Quality: By virtue of the physical dimensions of tubes or etched plate reactors, the surface area-to-volume ratio (SAVR) of a flow reactor is generally much higher than a typical batch reactor, meaning that heat transfer and mixing is superior in the flow process (note that the extent of mixing depends heavily on the nature of the flow,

such as laminar vs. turbulent flow) [34]. Generally, having reached an equilibrium state, flow chemistry then ensures consistent product quality: for example, purity and particle size (in the case of crystallization or nanoparticle formation) compared to batch processing, which suffers from batch-to-batch variability [33].
Safety: Compared to a batch process, the smaller reaction volume at any one time decreases the overall hazards of a flow process [28]. Moreover, the enhanced SAVR of a flow reactor allows for excellent control of exotherms, which might prevent the initiation of thermal runaways [35].
Optimization: Flow processes benefit from more rapid optimization than batch processes, owing to the ability to screen a variety of reaction conditions in shorter times, using a single reactor and one which is straightforward to automate [32, 36–38]. The higher SAVR enables faster changes in temperature. The residence time R_T (i.e., how long the reaction mixture is exposed to the reacting conditions; R_T = reactor volume/flow rate) can be varied simply by changing the flow rate and using the same input reaction mixture. Through the use of sample loops or the straightforward integration of process analytical technologies (PATs), reaction sampling can be expedited in flow processes and may give a more representative picture than sampling the bulk of a large batch reactor, while additionally minimizing exposure to the process operator [39]. Finally, the synergy of PATs and flow chemistry lends itself to the automation of processes [32, 38].

Notwithstanding the above, a fundamental challenge in flow chemistry is the processing of heterogeneous slurries that can cause irregular flow rates, or otherwise block the reactor/pump manifold [40]. In a general sense, flow chemistry often involves reactor design to fit a given chemical process. While this may not be an issue for certain industries that use simple chemical reactions to produce high-volume products (e.g., petrochemicals, fine chemicals), it does present issues to industries where companies require production of a portfolio of products using more complex, multistep chemistry (such as, e.g., the pharmaceutical and material industries) [32] where the design of a bespoke flow reactor for each given process may not be justifiable (cost-effective) in comparison with the versatility of a batch reactor used for multiple processes. In addition, since flow chemistry necessitates an interface of chemistry with engineering, and since chemists are traditionally taught batch processing methods in their education, a cultural barrier is sometimes presented in the uptake of flow chemistry in chemical industries. Additional training is typically required!

4.1.4 Synergy of Flow Chemistry and Microwave Heating

Microwave irradiation offers several benefits to high-temperature flow chemistry. The nature of conventional, conductive heating in flow chemistry involves thermal gradients across a reactor cross section (tube or channel). The central part of the tube will be cooler than the surfaces to which the conductive heating is directly applied (Fig. 4.1). A laminar flow regime will exacerbate these thermal gradients because

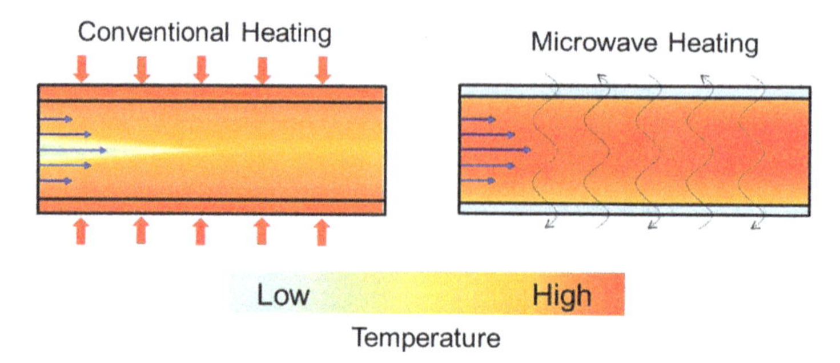

Fig. 4.1 Schematic illustration of conventional versus microwave heating temperature distribution (single-mode, resonant-type) in flow chemistry

of the effective increase in fluid velocity at the central part of the tube. In addition, the effective decrease in fluid velocity near the edges of the tube, to which higher temperature is applied, can lead to thermal decomposition or less desirable reaction profiles. The dissipation of thermal gradients as the reaction mixture moves across the conductively-heated tube will depend on the flow regime and on the flow rate. On the other hand, microwave heating involves the direct transfer of energy by dielectric loss. This inverts the thermal gradient [41–45], heating more uniformly across the cross section of the reactor with small losses occurring at the tube surface by conduction. Of course, smaller pipe diameters which mitigate the issue of penetration depth are preferred, but microwave heating does offer great potential for larger diameter tubes for the scale-up of high-temperature flow chemistry and can decrease the risk of thermal decomposition of low-velocity material near the tube edges.

Strauss and coworkers [43, 46] and Chen et al. [47] were the first to report the combination of flow chemistry with microwave heating. A host of reports have emerged since then, using various types of reactor configurations [48–94]. Classically, microwave flow chemistry has employed magnetron oscillators and multimode cavities whereby heating unevenness can be mitigated to some extent by the effect of the flow regime on thermal gradients.

Despite the fact that single-mode resonator-type cavities are desirable for controlled, reproducible and even heating, the size of such a resonator is limited to ca. 12.2 cm (the wavelength corresponding to the microwave frequency of 2.45 GHz). Considering this as the maximum size of batch microwave single-mode resonator, the amount of chemical compound that can be synthesized would be unsuitable for a commercial scale. Incorporation of single-mode cavities with well-defined electromagnetic field distributions is facilitated in flow chemistry, since flow reactor productivity is decoupled from reactor size. While single-mode cavities are gaining prominence in microwave flow chemistry; unfortunately, most reports still rely on magnetron oscillators.

4.1.5 Merits of a Semiconductor Microwave Generator in Flow Chemistry

Microwave heating of chemical reactions using single-mode devices with magnetron oscillators has fundamental complexities as (i) the heating efficiency depends on the dielectric properties of the solvent (ε' and ε'') [1–3, 15] and (ii) both ε' and ε'' exhibits temperature dependence [1, 2, 15, 95], thus causing the applied frequency of the microwave source and the resonant frequency of the cavity to fall out of sync as the temperature changes. These complexities result in unpredictable heating and poor process control. Although multimode devices with magnetron oscillators can avoid these problems (by emitting a broad range of frequencies), their electric field distributions are poorly characterized and irreproducible. Moreover, the energy inefficiency of such devices and the risk of explosion/perforation are increased because most of the electromagnetic energy is reflected. In order to heat non-polar solvents with a small ε'' to the desired high temperatures, one must employ very high applied microwave power. The issues mentioned above are undesirable in the context of flow chemistry, where the reaction mixture must be heated in a controlled and rapid fashion to reach the desired reaction temperature before flowing out of the reactor. This could explain why there are limited microwave flow chemistry reports utilizing non-polar solvents [96]. On the other hand, single-mode devices employing solid-state (semiconductor) microwave oscillators offer excellent process robustness, safety, and energy efficiency; compared to single-mode or multimode devices employing magnetron oscillators (Table 4.1).

Through the use of a resonant single-mode cavity, it is possible to uniformly irradiate such a flow channel with microwaves, which mitigates heating unevenness. Uniform heating of the flow path is critical in ensuring consistent quality in the synthesis of pharmaceuticals and fine chemicals. Moreover, where the oscillation frequency of the magnetron changes under the influence of the load, the solid-state generator can specify a single microwave oscillation frequency with high reproducibility. Parameters such as frequency and Q factor, as well as the microwave output power and reflected power, can be specified and are considered practical in terms of Good Manufacturing Practice (GMP). The longer life expectancy of solid-state devices compared to magnetrons is advantageous concerning the continuous

Table 4.1 Benefits of combining solid-state microwave single-mode devices with flow chemistry

	Benefit	Reason
1	Uniform heating	Defined and controlled electromagnetic field distribution (e.g., transverse electric (TE) mode, transverse magnetic (TM) mode)
2	Reproducibility	Defined, controllable and reproducible output
3	Energy efficiency	All applied power allocated to a single emitted frequency
4	Process reliability/Reliability	Longer life expectancy

operation of a flow reactor over a long period of time. Finally, the higher energy efficiency is desirable to process chemists.

As will be seen in the next section, by selecting the appropriate combination of cavity, waveguide, reactor vessel and pumping unit, a solid-state oscillator single-mode microwave flow reactor (i) can achieve large-scale synthesis, (ii) can heat non-polar solvents, (iii) can constantly apply microwave irradiation in a controlled manner to achieve controlled and even heating and (iv) offers a range of demonstrable opportunities to expedite reaction optimization. It is noted that in order to take full advantage of solid-state oscillators, device configuration and microwave control is paramount. If the key frequency or phase control features of the solid-state oscillator are not employed, a magnetron can be superior even for flow synthesis.

4.2 Semiconductor Generators: Microwave Flow Chemistry Applications

4.2.1 Reported Reactor Configurations and Capabilities

To our knowledge, the first example of a tunable solid-state (semiconductor) oscillator single-mode microwave flow reactor for chemical synthesis was disclosed by Nishioka et al. [85–87] for the synthesis of nanomaterials. Here, a quartz straight pipe was placed in a cylindrical single-mode cavity (Fig. 4.2) [88]. By this method, it was shown that Cu nanoparticles can be synthesized by the reduction of $Cu(OAc)_2$ by polyvinylpyrrolidone (PVP) in a short time compared with the batch reaction.

A tunable, solid-state oscillator single-mode microwave flow reactor developed by SAIDA FDS Inc. has been commercialized for organic synthesis applications (Fig. 4.3) [90–94, 97, 98]. This device continuously monitors and automatically

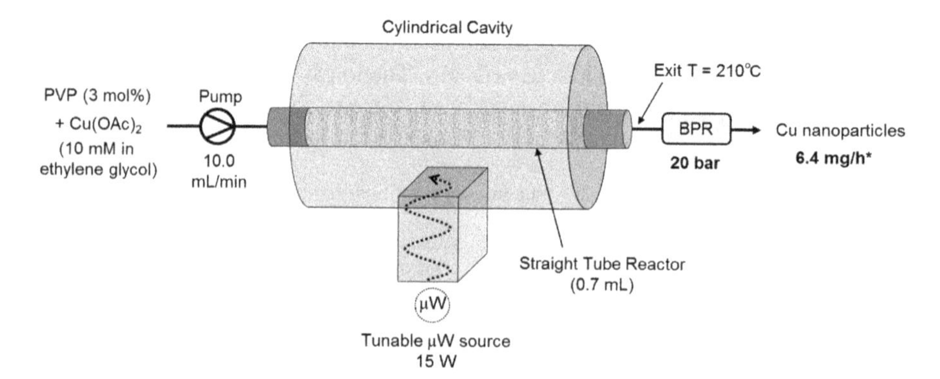

Fig. 4.2 Schematic of a bespoke solid-state oscillator single-mode straight tube microwave flow reactor developed by Nishioka et al. used for Cu nanoparticle synthesis. *In order to calculate the productivity, a 100% yield was assumed. Reproduced from Ref. [88]

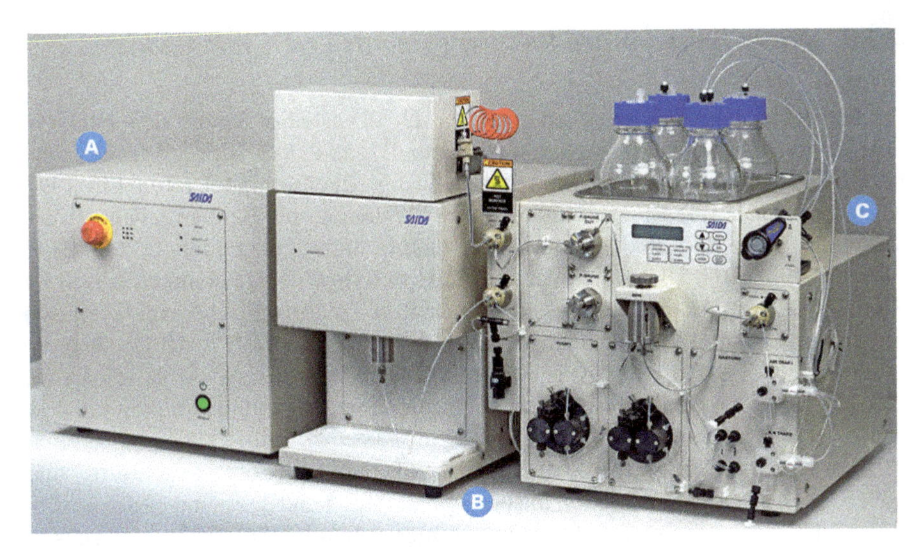

Fig. 4.3 Tunable, solid-state oscillator single-mode microwave flow reactor commercialized by SAIDA FDS Inc., **a** consisting of power supply unit, **b** microwave applicator unit and **c** pumping unit with two pumps

adjusts the applied microwave frequency to maximize the power of the electric field. Not only does this allow non-polar solvents to be rapidly heated, it also compensates for the changes of ε' and ε'' with temperature, thus resulting in excellent temperature and process control. The system can employ either a straight tube reactor (volume ca. 1.0 mL) or a helical tube reactor (volume ca. 1.0–6.0 mL).

The helical tube reactor presents several merits in contrast to the straight tube reactor. Firstly, it minimizes the resonant frequency variation for a number of solvents with different dielectric properties (from toluene to water) to that which remains approximately within the ISM band. Secondly, the helical shape permits a larger reaction volume to be placed within the central part of the cavity with maximized electric field power, thus increasing the available residence time. Finally, the secondary flows are promoted by the helical shape, resulting in increased radial mixing and turbulence inhibition compared to the straight tube reactor [92].

The system displayed in Fig. 4.3 has two pumps that, operated together, can deliver up to 2 × 9.9 mL min^{-1} (higher throughput systems capable of 2 × 50.0 mL min^{-1} are also available). Thanks to the rapid heat transfer afforded by microwave irradiation at such high flow rates, this system has been successfully employed for g hr^{-1} to hundreds of g hr^{-1} productivities of bulk chemicals, pharmaceutically relevant scaffolds and functional materials. Applications of tunable single-mode microwave flow using a solid-state oscillator are now presented.

4.2.2 *High-Temperature Rearrangements and Cycloadditions*

Fischer indole reactions are often catalyzed by Brønsted or Lewis acid catalysts. Often times, substrates do not tolerate acidic conditions and acid catalysts can present hazards to the user, to the environment and can suffer from poor reusability [99]. Therefore, uncatalyzed Fischer indole reactions driven only by thermal energy are sometimes more desirable, although they require high temperatures (>200 °C) to proceed [100, 101]. Using the microwave-assisted flow device commercialized by SAIDA FDS Inc., Akai et al. achieved an uncatalyzed thermal Fischer indole reaction of cyclohexanone (**1**) and phenylhydrazine (**2**) in AcOH/MeCN to afford (**3**) in up to 75% yield at 115.0 g hr^{-1} (Fig. 4.4a) [90]. Such productivity surpassed (by ca. 50 times) that of a similar Fischer indole reaction in a single-mode microwave flow device employing a magnetron oscillator [66].

Akai et al. found that the yield of **3** and the exit temperature of the reaction mixture increased when the concentration of starting materials **1** and **2** was increased, ascribing this behavior to selective absorption of microwaves by the starting materials or by the reactive intermediate **4** at those higher concentrations. While a conventionally

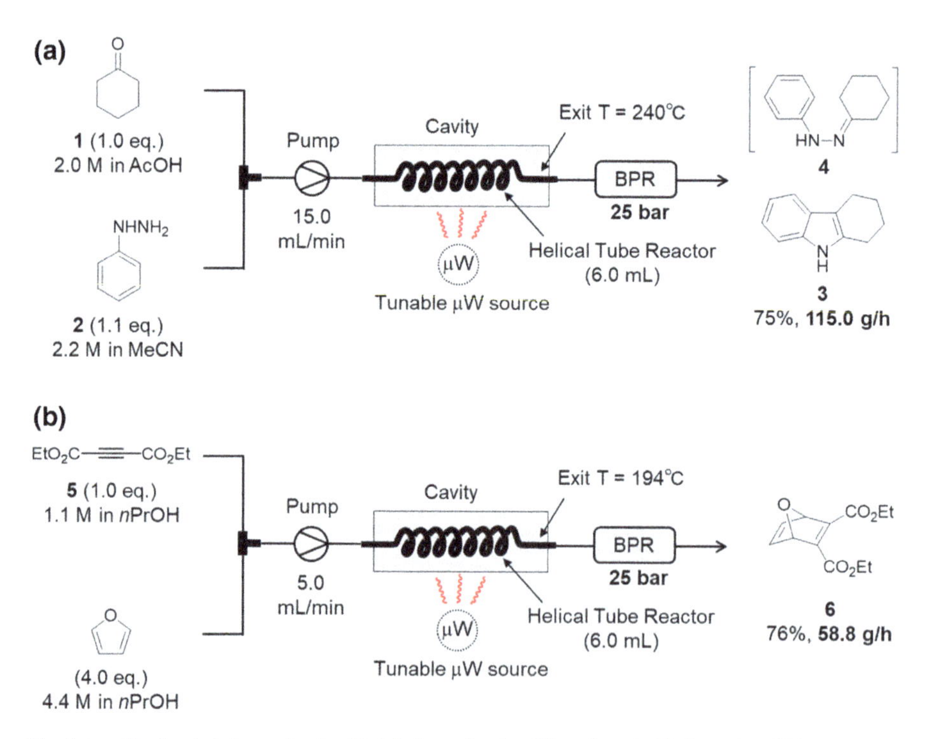

Fig. 4.4 **a** Fischer indole synthesis of 1,2,3,4-tetrahydro-1H-carbazole (**3**) from **1** and **2** in a tunable, solid-state oscillator single-mode microwave flow reactor. **b** Diels–Alder reaction of furan with diethyl acetylenedicarboxylate (**5**), to afford compound **6** in a tunable, solid-state oscillator single-mode microwave flow reactor. Reproduced from Ref. [90]

heated control reaction was not reported in this study, superior results afforded by microwave heating compared to conventional heating are reported in related studies [100, 101]. A recent HCl-catalyzed Fischer indole reaction reported that notably higher (by ca. 20%) yields resulted from microwave flow (single-mode, using a magnetron) in comparison with conventionally heated flow [102].

Kinetic studies allowed the authors [90] to identify local enhanced temperatures as responsible for accelerating the [3+3] sigmatropic rearrangement. Moreover, Akai et al. demonstrated the microwave-assisted flow Diels–Alder of diethyl acetylenedicarboxylate **5** with furan using a modified reactor setup, affording product **6** in up to 76% yield at 58.8 g hr^{-1} (Fig. 4.4b) [90]. Compared to similar microwave-assisted flow Diels–Alder reactions, this represented the highest reported productivity [103, 104].

A tunable, solid-state oscillator single-mode microwave flow reactor was applied to the Wolff rearrangement of α-diazoketone **7** by Ley et al. [91] (Fig. 4.5a). The ketene **8** was formed in situ, then trapped with benzylamine **9** in MeCN to give up to 99% yield at 0.34 g hr^{-1}. Where sealed batch vessels under microwave heating or stainless-steel coil flow reactors under conventionally heating were employed, the

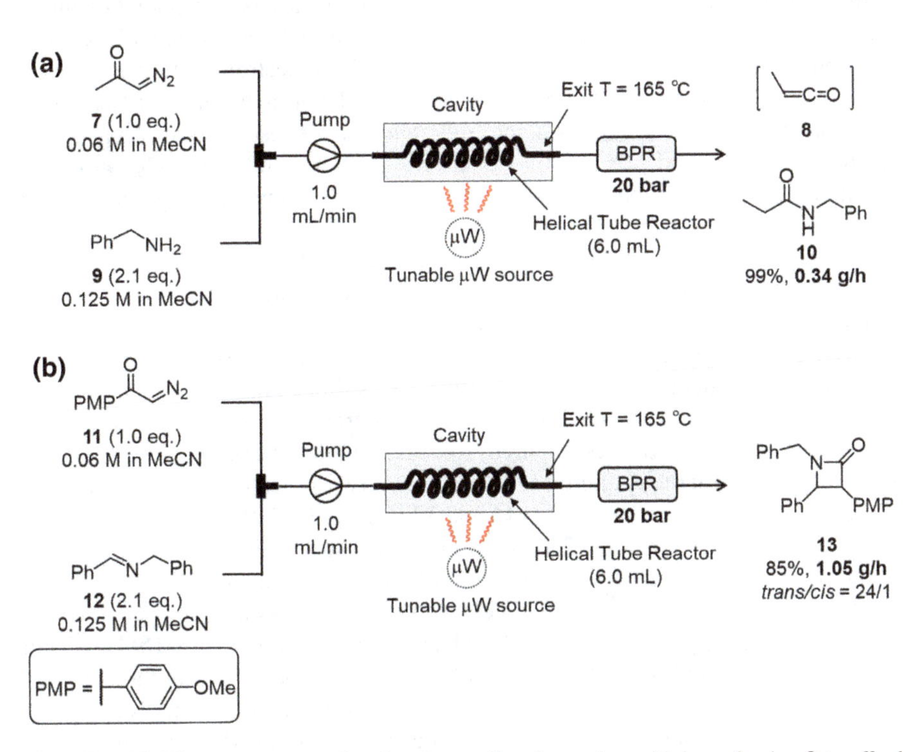

Fig. 4.5 **a** Wolff rearrangement of α-diazoketone **7** and trapping with benzylamine **9** to afford benzylamide **10** in a tunable, solid-state oscillator single-mode microwave flow reactor. **b** Synthesis of β-lactam **13** by Wolff-Staudinger reaction between α-diazoketone **11** and imine **12** in a tunable, solid-state oscillator single-mode microwave flow reactor. Reproduced from Ref. [91]

authors reported inferior yields for this reaction. Moreover, in order to extend the reaction to the syntheses of β-lactams such as **13**, the ketene intermediates resulting from the Wolff rearrangement of α-diazoketones such as **11** were intercepted in Staudinger-type [2+2] cycloadditions with imines such as **12** (Fig. 4.5b). This method gave β-lactam **13** in up to 85% yield at 1.05 g hr^{-1} and is important because despite the value of its final β-lactam products, the Wolff–Staudinger reaction is a transformation known to present problems during scale-up. High temperatures are required, gaseous N_2 is released from the Wolff rearrangement step and α-diazocarbonyl compounds are hazardous, potentially explosive compounds [105–107]. Notably, Ley et al. [91] found that the high temperature reached in the microwave flow reactor was crucial in the *trans/cis* diastereoselectivity of lactam formation (the reaction mechanism was investigated computationally, employing DFT calculations at high temperature conditions to rationalize the observed diastereoselectivities). Despite the modest productivities (ca. 0.3–1.0 lactam), this report exemplifies the valuable application of a tunable solid-state oscillator single-mode microwave flow system to the safe generation and reaction of hazardous reactive intermediates at high temperatures in a fully contained environment.

In the Diels–Alder reaction of **5** with furan (Fig. 4.4b), an electron-poor dienophile reacts with an electron-rich diene. Diels–Alder reactions with 'mismatched' electronics, such as an electron-rich dienophile reacting with an electron-rich diene, require more forcing conditions of temperature and time. Upon thermal tautomerization of indene **15** to an electron-rich diene, fullerene C_{60} **14** reacts as an electron-rich dienophile to afford indene C_{60} monoadduct ($IC_{60}MA$, **16**) and indene C_{60} bisadduct ($IC_{60}BA$, **17**); see Fig. 4.6a. In batch mode, under conventional heating, the reaction requires several days of heating at $\geq$180 °C [108, 109].

Fullerenes exhibit poor solubility in most organic solvents; in fact, their solubility decreases with increasing temperature. Therefore, dilute conditions must be employed if reactions are to be scaled up via flow chemistry because of the risk of precipitation at high temperatures that would lead to clogging of the reactor. Consequently, the productivity of materials **16** and **17** suffers. High flow rates could be envisaged as an offset, provided that efficient heat transfer can be accomplished to achieve the high temperatures required. Recognizing this, Barham, Norikane et al. capitalized on the merits of microwave flow chemistry in order to bolster the synthesis of organic photovoltaic components **16** and **17** (Fig. 4.6b) [92].

Yields of up to 57% and 32% of **16** and **17**, respectively, were achieved and conditions were identified for their gram/hour scale production (0.74 g hr^{-1}, a 46% yield of **16** and 0.47 g hr^{-1}, a 32% yield of **17**, respectively). Such productivities surpassed previously reported methods in batch mode (0.04 g hr^{-1} of **16**) [110] and conventionally heated flow (0.07 g hr^{-1} of **17**) [111]. In order to minimize the use of chlorinated solvents and improve the 'greenness' of the chemistry, commonly used *o*-DCB was replaced with *o*-xylene in this study and this did not compromise the yield or productivity. However, since the permittivity of *o*-xylene is significantly lower than *o*-DCB ($\varepsilon = 2.6$ vs. 9.9, respectively), the rapid and controlled heating of reactions would present a challenge to multimode or non-tunable single-mode microwave flow reactors. By use of SAIDA FDS Inc.'s single-mode device with a

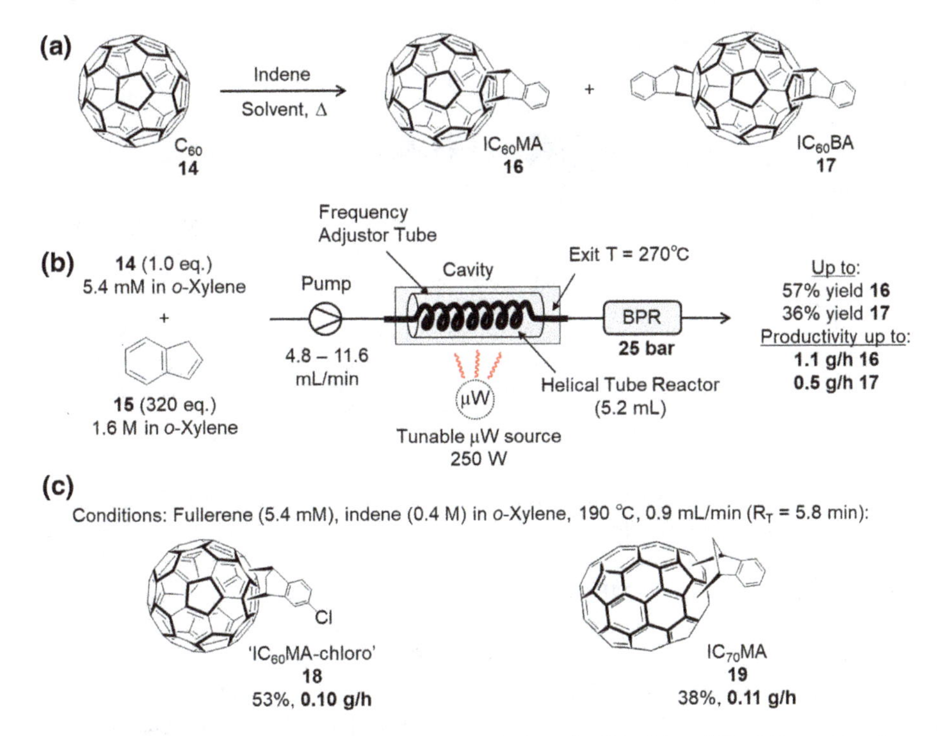

Fig. 4.6 **a** Indene-C_{60} Monoadduct ($IC_{60}MA$) and Indene-C_{60} Bisadduct ($IC_{60}BA$) resulting from a high temperature Diels–Alder reaction of C_{60} with indene. **b** Fullerene/indene monoadduct and bisadduct syntheses in a tunable, single-mode microwave applicator. **c** Syntheses of '$IC_{60}MA$-chloro' and $IC_{70}MA$ using a tunable solid-state oscillator single-mode microwave flow reactor. Reproduced from Ref. [92]

solid-state oscillator, auto-frequency tunable microwave heating successfully heated *o*-xylene to 270 °C in minutes. Under decreased productivity (sub-g hr^{-1}) conditions, the synthesis of novel derivative '$IC_{60}MA$-chloro' (**18**) and $IC_{70}MA$ (**19**) without prior batch studies demonstrated the robustness of the process (Fig. 4.6c). Benefits of tunable, solid-state oscillator single-mode microwave-heated flow were observed during reaction optimization; see Sect. 4.2.5.

Permittivity data measurements gave identical results for the reaction mixture and solvent *o*-xylene, suggesting that 'microwave effects' were not operational in this process. Indeed, a comparison reaction using conventional heating gave identical product yields, conversion and selectivity. Nonetheless, the superior heat transfer of microwave heating is anticipated to exhibit a benefit in the further scale-up of this reaction if larger diameter tube reactors and higher flow rates were to be used.

Koyama et al. [112] investigated the Claisen rearrangement reaction of allyl 1-naphthyl ether **20** in a tunable single-mode microwave flow reactor to give 2-allylnapthalen-1-ol **21** in up to 96% yield at 20.3 g hr^{-1} productivity (Fig. 4.7). Notably, the rapid heating of non-polar solvent CPME to ~100 °C above its boiling point (achieved in ca. 10 min at 45 W applied microwave power, 1.0 mL min^{-1}

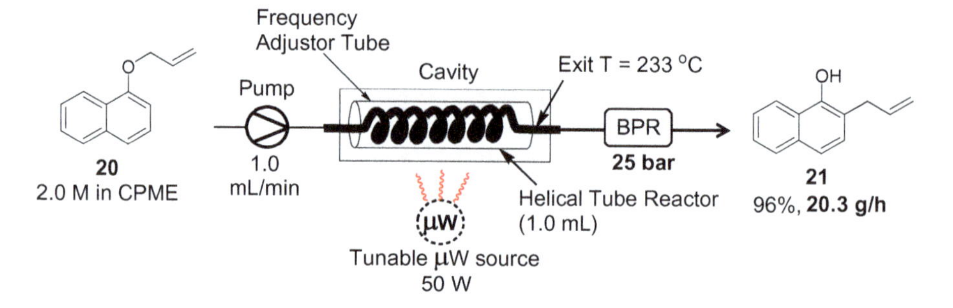

Fig. 4.7 Synthesis of 2-allylnapthalen-1-ol **21** by aromatic Claisen rearrangement of allyl 1-naphthyl ether **20** in a tunable, solid-state oscillator single-mode microwave flow reactor. Reproduced from Ref. [112]

flow rate) demonstrates the capability of the tunable solid-state oscillator single-mode microwave flow reactor and the safety benefits derived from flow chemistry. An increase in reactor exit temperature and the yield of **21** was observed when the concentration of **20** in CPME was increased while holding other reaction conditions (flow rate, applied microwave power, back pressure) constant. Moreover, tan δ increased with increasing concentration of **20** (0.1 M, tan $\delta = 0.042$; 1.0 M, tan $\delta = 0.062$; 2.0 M, tan $\delta = 0.085$), a phenomenon that can be ascribed to selective substrate heating in the presence of solvent. Although a conventionally heated flow comparison was not conducted in this study [112], the apparent selective substrate heating is expected to derive higher product yields from microwave-heated flow than conventionally heated flow; the authors envisaged the benefits of this enhanced heat transfer toward large-scale operations.

Hamashima et al. reported the first example of a Johnson–Claisen rearrangement in a continuous flow microwave system (Fig. 4.8) [113]. Allyl alcohol **22** and triethylorthoacetate **23** in the presence of AcOH reacted to give the γ,δ-unsaturated ester **24** in up to 87% yield at 84.6 g hr^{-1} productivity. The robustness of the process was demonstrated by synthesis of a selection of derivatives **25–29** that were obtained in good to excellent (50–88%) yields. Interestingly, when the substrates were changed and other reaction conditions held constant (for example, comparing syntheses of **28** vs. **29**), the exit temperature of the reactor differed. This suggests selective heating of the allylic alcohol or its derived transition state. The efficiency of microwave heating in this reaction was found to be heavily influenced by the loading of the acid catalyst (while holding other conditions constant; increasing AcOH from 0.05 to 0.5 eq. resulted in a ca. 20 °C decrease in steady-state temperature). The varying amounts of AcOH were claimed to alter the microwave absorption characteristics of the components of the reaction mixture, although the exact role of AcOH remains unclear.

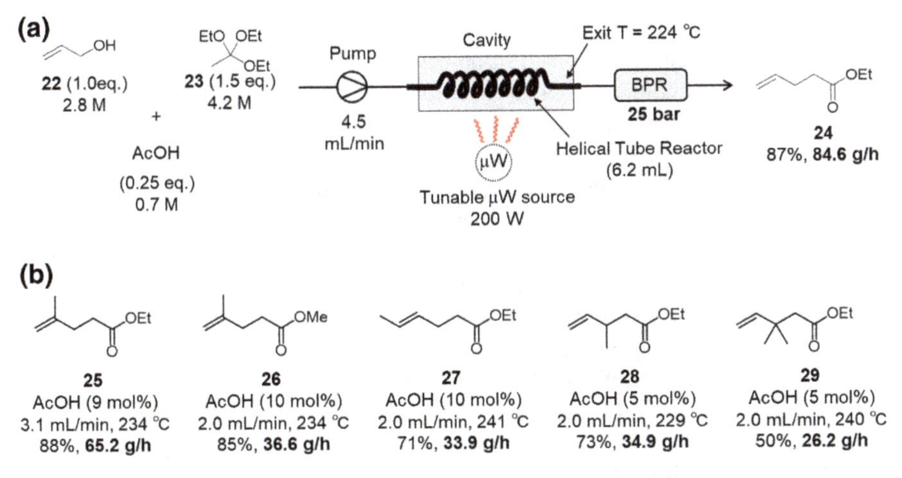

Fig. 4.8 **a** Johnson-Claisen reaction of allyl alcohol **22** and triethylorthoacetate **23** providing γ,δ-unsaturated ester **24** in a tunable, solid-state oscillator single-mode microwave flow reactor. **b** Substrate scope of the microwave-assisted Johnson-Claisen rearrangement. Reproduced from Ref. [113]

4.2.3 *High-Temperature Alkylation Reactions*

A Williamson etherification reaction using the microwave-assisted flow device commercialized by SAIDA FDS Inc. was recently disclosed by Hamashima et al. [94] (Fig. 4.9). Here, 4-hydroxycinnamic acid **25** and 5-bromo-1-pentene **26** were reacted together with potassium hydroxide as a base at 160 °C. This afforded (*E*)-3-[4-(pent-4-en-1-yloxy)phenyl]acetic acid **27**, a component of liquid crystal materials, in up to 91% yield (isolated) at 12.7 g hr^{-1}. The previously reported batch synthesis of **27** in EtOH/H_2O (3:1) at reflux over 24 h represented a theoretical productivity of 92.9 mg hr^{-1} [114]. A MeOH/H_2O solution was employed after the reactor exit and

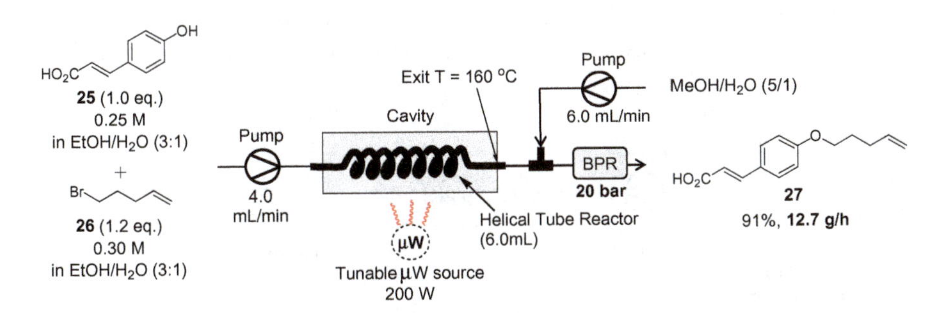

Fig. 4.9 Synthesis of (*E*)-3-[4-(pent-4-en-1-yloxy)phenyl]acetic acid **27** from 4-hydroxycinnamic acid **25** and 5-bromo-1-pentene **26** in a tunable, solid-state oscillator single-mode microwave flow reactor. Reproduced from Ref. [94]

before the back-pressure regulator in order to prevent blocking of the reaction by precipitation of the carboxylate salt of product **27**.

The authors [94] compared batch reactions using microwave heating to conventionally heated reactions (using an oil batch) under the same conditions of temperature (90 °C). After the same reaction time, higher yields resulted from the batch microwave reactions. However, the reliable monitoring of temperature is critical for investigating 'microwave effects' and batch mode microwave reactions as they oftentimes suffer from localized temperature gradients that result in temperatures very different from the 'average' temperature displayed by the reactor's temperature probe. Therefore, rate enhancements may simply be attributed to elevated local reaction temperatures [115]. If controlled, single-mode microwave heating were employed, comparisons in flow would be more useful in this respect, because the flowing reaction mixture would result in a comparable temperature gradient across the reactor for each mode of heating (for a given reactor configuration).

Barham, Hamashima et al. [116] reported an alkoxide base (NaO*t*Bu/KO*t*Bu)-mediated C-alkylation of amides with styrenes, where scalability of the reaction of *N*-methylpyrrolidone with styrene **28** using a microwave flow reactor was investigated (Fig. 4.10). Using NaO*t*Bu (0.6 eq.) as base, 18-crown-6 (0.6 eq.) as a solubilizing and activating additive and heating at 180 °C, product **29** was afforded in 73% yield (yield determined by ^{1}H NMR, the isolated yield was 70%) at 64.9 g hr^{-1} (calculated on the basis of the ^{1}H NMR yield). Operating the reaction under the boiling point of *N*-methylpyrrolidone meant that the BPR (whose fine channels were prone to blocking) was not necessary.

The reaction (C-alkylation of *N*-methylpyrrolidine with styrene) was conducted under milder conditions (that did not allow reaction completion) to compare heating from the microwave versus heating from a conventional heater (furnace heater). In the control reaction, no reaction was observed after processing the freshly prepared reaction mixture in the absence of heat (Fig. 4.11) [117]. For the microwave and conventional heating comparison, the same exit temperature (140 °C) and residence time ($R_T = 0.55$ min) were ensured. Microwave heating approximately doubled the yield of **29** (30%) when compared with conventional heating (14%). This observation contrasts with the identical yields and selectivities observed when comparing microwave-heated and conventionally-heated flow reactions of fullerene and indene

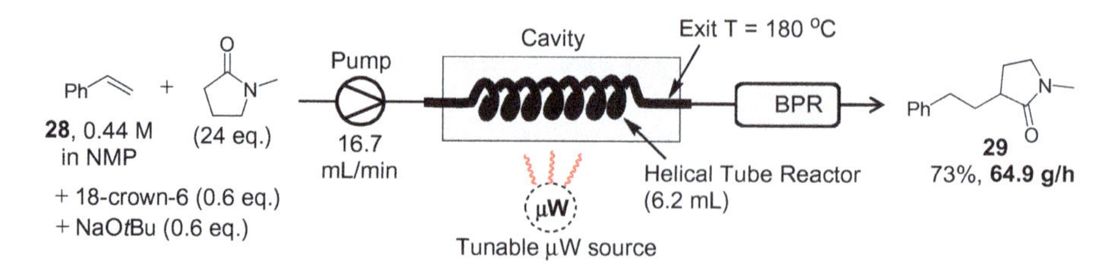

Fig. 4.10 Microwave flow C-alkylation of *N*-methylpyrrolidine with styrene **28** to give product **29** in a tunable, solid-state oscillator single-mode microwave flow reactor. Reproduced from Ref. [116]

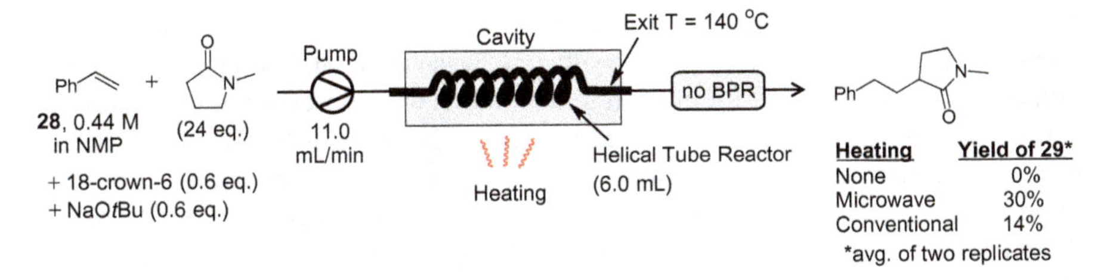

Fig. 4.11 Tunable, solid-state oscillator single-mode microwave-heated flow versus conventionally heated flow C-alkylation of *N*-methylpyrrolidine with styrene **28**

mentioned earlier [92]. The reaction mixture and solvent had similar tan δ values at room temperature (rt), 2.45 GHz (NMP only, tan $\delta = 0.888$ vs. NMP + styrene + 18-crown-6 + NaO*t*Bu, tan $\delta = 0.821$) evidencing against non-thermal 'microwave effects' (selective transfer of energy to the reagents) [118–120]. More likely is the case that microwave heating enables more effective energy transfer to the fast-flowing mixture over the same given residence time than conventional heating does (the permittivity value at high temperature may differ from those at rt so a non-thermal 'microwave effect' cannot be completely ruled out).

4.2.4 Heterogeneous Catalytic Reactions

A Mizoroki–Heck reaction under heterogeneous catalysis was explored by Sajiki et al. [93] using the microwave-assisted flow device commercialized by SAIDA FDS Inc. (Fig. 4.12). A straight tube reactor fitted with a glass filter was packed with *N,N*-dimethylamino-functionalized anion-exchange resin-embedded Pd (7% Pd/WA30) [121] as the catalyst and housed within the microwave cavity. By flowing a mixture of 4-iodoacetophenone **30** (0.25 M), *n*-butylacrylate (**31**) and *n*-tributylamine in MeCN through the catalyst cartridge at 80–100 °C, the authors [93] achieved the synthesis of **32** in up to 91% yield at 0.50 g hr^{-1}. Atomic absorption analysis demonstrated that Pd leaching into the reaction mixture was negligible, thereby rationalizing the ability to

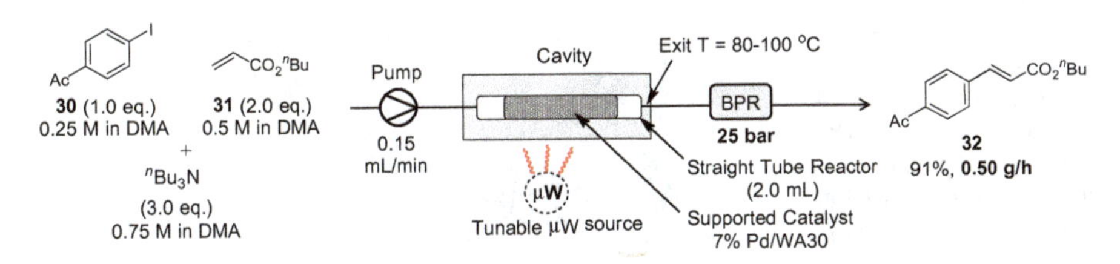

Fig. 4.12 Mizoroki–Heck reaction of 4-iodoacetophenone **30** and *n*-butylacrylate **31** to afford product **33** in a tunable, solid-state oscillator single-mode microwave flow reactor containing a Pd/WA30-packed straight tube. Reproduced from Ref. [93]

reuse the catalyst cartridge up to 5 times without significant change in conversion or yield of product. When the aryl iodide substrate was varied while holding other reaction conditions constant, different exit temperatures were observed despite the same applied microwave power. The authors ascribed this behavior to selective substrate heating of certain substrates. A modest improvement in the conversion and product yield for the microwave-assisted flow device compared to a conventionally heated flow reactor was also reported and rationalized by the more rapid and efficient heat transfer of microwave heating. To our knowledge, this was the first report of a flow reaction under heterogeneous catalysis which takes place under microwave-heating using a tunable solid-state oscillator.

4.2.5 Reaction Optimization

The robustness offered by tunable (solid-state oscillator) single-mode microwave flow has been leveraged toward reaction optimization. Hamashima et al. [113] used a one-factor-at-a-time (OFAT) method to find optimum reaction conditions for their microwave-assisted flow Johnson–Claisen rearrangement reaction (Fig. 4.8). Here, a response surface was constructed based on the experimental data points which were obtained by varying the flow rate and additive (AcOH) equivalents (Fig. 4.13). Mase et al. [37] used the '9 + 4 + 1 method' to optimize various reactions (alcohol acetylation, a Fischer indole syntheses, a Diels–Alder reactions). The '9 + 4 + 1 method' is another type of OFAT method that involves extrapolating a surface approximation for result prediction (Fig. 4.14). Thus, the upper and lower limits of temperature and flow rate were divided into three (9 experiments) and 4 more experiments were carried out about the optimum of these initial 9 points. A final experiment (1) was carried out to validate the expected optimal conditions (14 experiments in total). Both OFAT approaches are reasonable approaches to improve reaction conditions where

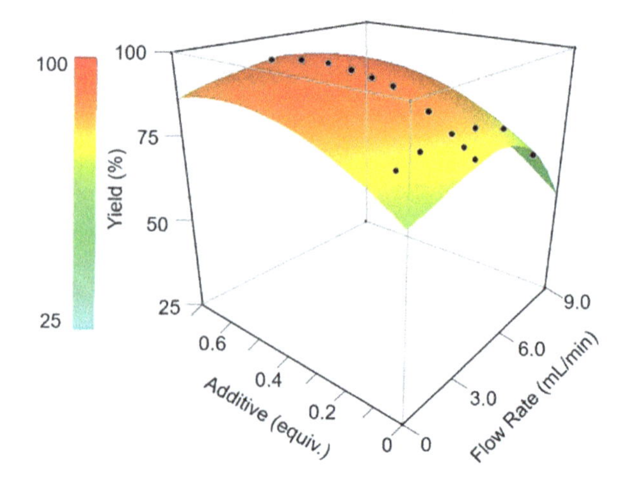

Fig. 4.13 Illustration of a response surface constructed from a one-factor-at-a-time (OFAT) design applied to a Johnson-Claisen rearrangement investigated by tunable (solid-state oscillator) single-mode microwave flow. *Note* Data differ from the reported studies and are shown for illustrative purposes only

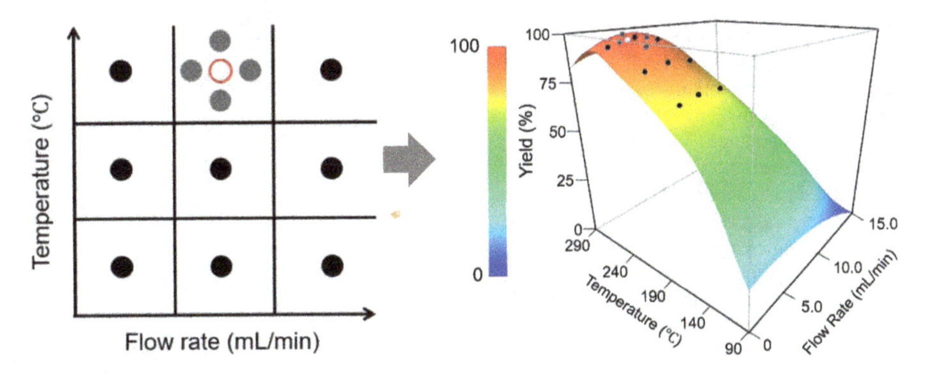

Fig. 4.14 Illustration of a '9 + 4 + 1 OFAT method' and design applied to various reactions investigated by tunable (solid-state oscillator) single-mode microwave flow. *Note* Data differs from the reported study and is shown for illustrative purposes only

a strong starting point has already been identified, while the '9 + 4 + 1 method' has the potential to minimize the number of experiments required to generate the surface.

In the Design of Experiment analysis [122], factorial designs are generally more efficient compared to the previously mentioned OFAT methods, which provide similar or more information for a similar experimental cost. Barham, Norikane et al. [92] utilized the two-factorial face-centered central composite design (CCD), illustrated in Figs. 4.15 and 4.16, for their investigations of a C_{60}/fullerene-indene Diels–Alder reaction (Fig. 4.6) [92]. Theoretically, a CCD requires 15 experiments in total (8 experiments for the 'cube' design, 6 experiments for the 'star' design and a central point experiment. Practically, several central point experiments are required to statistically validate the model (5 central point experiments were used by Barham, Norikane et al. [92]). This statistically and experimentally verified model allowed a deep understanding of the relative importance of factors affecting C_{60} conversion (temperature, [indene], residence time) and accurately predicted results within and around the design space (Fig. 4.16). Because of the factorial nature of its design, the

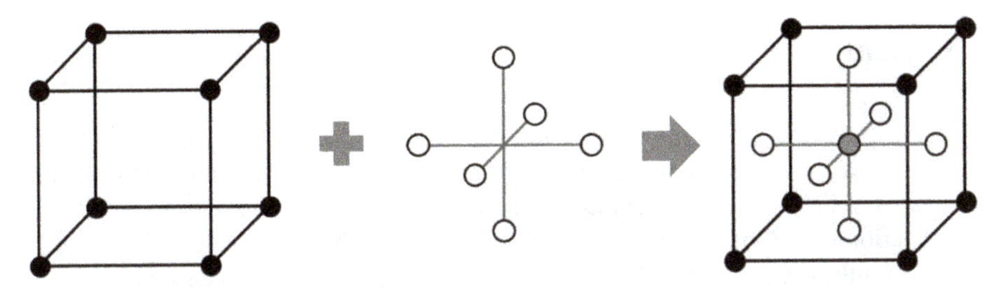

Fig. 4.15 Conceptual combination of two-level full factorial cubic design and face-centered star design to afford a two-factorial face-centered central composite design (CCD)

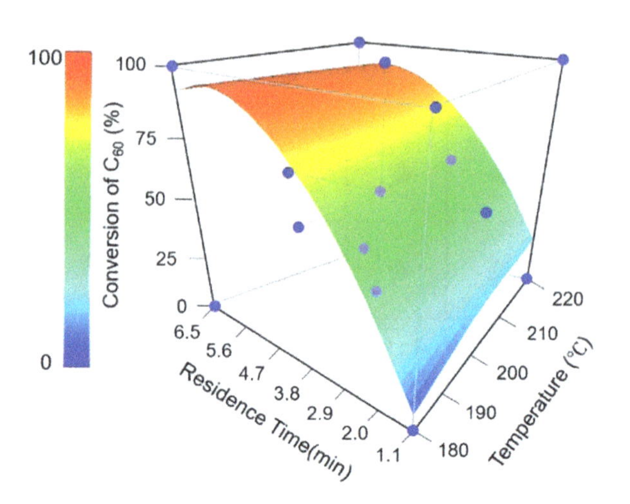

Fig. 4.16 Illustration of two-factorial face-centered central composite design (CCD) for a C_{60}/fullerene-indene Diels–Alder reaction investigated by tunable (solid-state oscillator) single-mode microwave flow. *Note* Data differs from the reported study and is shown for illustrative purposes only. Reproduced from Ref. [92]

face-centered CCD model can reveal multifactor interactions (an ability which distinguishes it from OFAT models), although no significant interactions were detected in the C_{60}/fullerene-indene Diels–Alder reaction. Since the face-centered CCD required reaction testing under 'extreme' conditions (conditions dictated by the corners of the predetermined design space), rather than incremental variations around a single factor or constricted surface of design space, it presented an excellent challenge to the robustness of tunable (solid-state oscillator) single-mode microwave flow technology.

In the construction of all these models, the ability to rapidly access new, stable temperatures via microwave heating and flow processing lent itself to rapid data acquisition (<5 min interval between runs, on average, for the face-centered CCD illustrated in Fig. 4.16) [92]. In this regard, the ability of tunable (solid-state oscillator) single-mode microwave heating to continuously detect and match the temperature-dependent microwave absorption frequency allows for controlled changes in temperature, which is key to rapidly achieving new stable temperatures.

4.3 Summary and Outlook

Sections 4.1.1 and 4.1.2 introduced the roles of microwave heating in organic synthesis, the different approaches to microwave heating device configurations, and compared solid-state oscillators and magnetron oscillators as microwave generators. Sections 4.1.3–4.1.5 introduced the principles and benefits of flow chemistry in organic synthesis, the merits that derive from the synergy of microwave heating and flow chemistry and the benefits of solid-state oscillators in organic synthesis and, particularly, in flow chemistry.

Section 4.2 presented applications of solid-state oscillators in flow chemistry. Section 4.2.1 introduced reported reactor configurations and capabilities, while Sects. 4.2.2–4.2.4 described examples of tunable (solid-state oscillator) single-mode microwave flow technology in organic synthesis. Attention was drawn to the benefits of microwave flow processes, in general, as well as the specific benefits of tunable single-mode microwave heating on those chemistries. These include (i) the ability to heat non-polar solvents, (ii) the harnessing of microwave thermal effects and (iii) the rapid nature of microwave heating coupled with flow processing that permits the rapid changing of reaction conditions. Tunable single-mode microwave flow was used to scale up reactions relevant to the synthesis of bulk chemicals, pharmaceuticals and functional materials as far as several to hundreds of g hr^{-1} productivities. Finally, several approaches to the reaction optimization of microwave flow reactions were introduced in Sect. 4.2.5. The ability to explore different experimental designs, including those with 'extreme' conditions, shows the practicability and robustness of tunable single-mode microwave heating in flow reactions. Moreover, the rapid acquisition of data to construct said designs was realized.

Overall, this chapter titled 'Microwave Flow Chemistry' imparts readers with an understanding and awareness of the benefits of tunable solid-state oscillator-generated single-mode microwave heating. This will empower chemists involved in large-scale production to recognize and incorporate solid-state (semiconductor) oscillator-generated microwave flow technology into their processes as a robust, controlled technology that can replace other modes of microwave chemistry, and can perform comparably or superior chemistry to conventional thermal flow chemistry.

Acknowledgements We thank Professors Y. Hamashima, H. Egami and Akai (University of Shizuoka), Professors H. Sajiki and Y. Monguchi (Gifu Pharmaceutical University), Professors N. Mase, K. Takeda and K. Sato (Shizuoka University), Professor S. V. Ley (University of Cambridge), Professors Y. Norikane and J. Sugiyama (National Institute for Advanced Industrial Science and Technology (AIST), Japan) for collaborations and helpful discussions. We thank Mr. T. Okamoto and Mr. H. Odajima at Pacific Microwave Technologies for assisting with resonator and reactor design. We thank Professors Y. Wada, E. Suzuki, S. Fujii, M. Maitani and S. Tsubaki (Tokyo Institute of Technology), Professor S. Mineki (Tokyo University of Science), and Professor S. Ohuchi (Kyushu Institute of Technology) for their supports. We are grateful for the financial support from the Subsidy Program for Innovative Business Promotion of Shizuoka Prefecture to support our collaborative work. Joshua P. Barham is a former JSPS International Research Fellow and is grateful for financial support from JSPS.

References and Notes

1. Stuerga D (2006) Microwaves in organic synthesis, 2nd edn, Loupy A (ed). Wiley-VCH, Weinheim
2. Gabriel C, Gabriel S, Grant EH, Halstead BSJ, Mingos DMP (1998) Dielectric parameters relevant to microwave dielectric heating. Chem Soc Rev 27:213–224
3. Bogdal D, Prociak A (2008) Microwave-enhanced polymer chemistry and technology. Wiley-VCH, Weinheim

4. See https://www.itu.int/en/Pages/default.aspx
5. Gulich R, Köhler M, Lunkenheimer P, Loidl A (2009) Dielectric spectroscopy on aqueous electrolytic solutions. Radiat Environ Biophys 48:107–114
6. Metaxas AC, Meredith RJ (2008) Industrial microwave heating, Johns AT, Ratcliff G, Platts JR (eds). Lightning Source UK Ltd, Milton Keynes
7. Sturm GSJ, Verweij MD, van Gerven T, Stankiewicz AI, Stefanidis GD (2012) On the effect of resonant microwave fields on temperature distribution in time and space. Int J Heat Mass Transf 55:3800–3811
8. Weeson R, Jerby E, Schwarz E, Gerling JF, Werner K, Durnan G, Yakovlev VV, Achkasov K, Meir Y, Metaxas AC (2016) Trends in RF and microwave heating: special issue on solid-state microwave heating. AMPERE Newslett. https://doi.org/10.3390/cryst8100379
9. Schwartz E, Anaton A, Huppert D, Jerby E (2006) Transistor-based miniature microwave heater. In: 40th annual microwave symposium proceedings. https://doi.org/10.1109/tmtt.2012.2198233
10. Horikoshi S, Schiffmann RF, Fukushima J, Serpone N (2018) Microwave chemical and materials processing. A tutorial. Springer Nature, Singapore. https://doi.org/10.1007/978-981-10-6466-1
11. Atuonwu JC, Tassou SA (2019) Energy issues in microwave food processing: a review of developments and the enabling potentials of solid-state power delivery. Crit Rev Food Sci Nutr 59(9): 1392–1407. https://doi.org/10.1080/10408398.2017.1408564
12. Atuonwu JC, Tassou SA (2018) Quality assurance in microwave food processing and the enabling potentials of solid-state power generators: a review. J Food Eng 234: 1–15. https://doi.org/10.1016/j.jfoodeng.2018.04.009
13. Ohtani T (1973) Hybrid microwave heating apparatus. New Nippon Electron Co. Ltd. US Patent No. US3867607A
14. Metaxas AC, Meredith RJ (1983) Industrial microwave heating. Peter Peregrinus Ltd, London
15. Chandrasekaran S, Ramanathan S, Basak T (2012) Microwave material processing—a review. AIChE J 58:330–363
16. Kappe CO, Stadler A, Dallinger D (2013) Microwaves in organic and medicinal chemistry, Mannhold R, Kubinyi H, Folkers G (eds), Chaps. 3.3–3.5. Wiley-VCH, Weinheim
17. Kurniawan H, Alapati S, Che WS (2015) Effect of mode stirrers in a multimode microwave-heating applicator with the conveyor belt. Int J Prec Eng Manuf 2:31–36
18. Ano T, Kishimoto F, Sasaki R, Tsubaki S, Maitani MM, Suzuki E, Wada Y (2016) In situ temperature measurements of reaction spaces under microwave irradiation using photoluminescent probes. Phys Chem Chem Phys 18:13173–13179
19. Razzaq T, Kremsner JM, Kappe CO (2008) Investigating the existence of nonthermal/specific microwave effects using silicon carbide heating elements as power modulators. J Org Chem 73:6321–6329
20. Gutmann B, Obermayer D, Reichart B, Prekodravac B, Irfan M, Kresmsner JM, Kappe CO (2010) Sintered silicon carbide: a new ceramic vessel material for microwave chemistry in single-mode reactors. Chem Eur J 16:12182–12194
21. Liu X, Zhang Z, Wu Y (2011) Absorption properties of carbon black/silicon carbide microwave absorbers. Compos B Eng 42:326–329
22. Kappe CO (2013) Unraveling the mysteries of microwave chemistry using silicon carbide reactor technology. Acc Chem Res 46:1579–1587
23. (a) Giguere RJ, Bray TL, Duncan SM, Majetich G (1986) Application of commercial microwave ovens to organic synthesis. Tetrahedron Lett 27(41):4945–4948; (b) Gedye R, Smith F, Westaway K, Ali H, Baldisera L, Laberge L, Rousell J (1986) The use of microwave ovens for rapid organic synthesis. Tetrahedron Lett 27(3):279–282
24. Strauss CR, Trainor RW (1995) Developments in microwave-assisted organic chemistry. Aust J Chem 48:1665–1692
25. Ngoc TL, Roberts BA, Strauss CR (2006) Microwaves in organic synthesis, 2nd edn, Loupy A (ed). Wiley-VCH, Weinheim

26. Devine WG, Leadbeater NE (2011) Probing the energy efficiency of microwave heating and continuous-flow conventional heating as tools for organic chemistry. Arkivoc 5:127–143
27. Roberge DM, Zimmermann B, Rainone F, Gottsponer M, Eyholzer M, Kockmann N (2008) Microreactor technology and continuous processes in the fine chemical and pharmaceutical industry: is the revolution underway? Org Process Res Dev 12:905–910
28. Gutmann B, Cantillo D, Kappe CO (2015) Continuous-flow technology—a tool for the safe manufacturing of active pharmaceutical ingredients. Angew Chem Int Ed 54:6688–6728
29. Cole KP, Groh JM, Johnson MD, Burcham CL, Campbell BM, Diseroad WD, Heller MR, Howell JR, Kallman NJ, Koenig TM, May SA, Miller RD, Mitchell D, Myers DP, Myers SS, Phillips JL, Polster CS, White TD, Cashman J, Hurley D, Moylan R, Sheehan P, Spencer RD, Desmond K, Desmond P, Gowran O (2017) Kilogram-scale prexasertib monolactate monohydrate synthesis under continuous-flow CGMP conditions. Science 356:1144–1150
30. Myers RM, Fitzpatrick DE, Turner RM, Ley SV (2014) Flow chemistry meets advanced functional materials. Chem Eur J 20:12348–12366
31. Moghaddam MM, Baghbanzadeh M, Sadeghpour A, Glatter O, Kappe CO (2013) Continuous-flow synthesis of CdSe quantum dots: a size-tunable and scalable approach. Chem Eur J 19:11629–11636
32. Jensen KF, Reizman BJ, Newman SG (2014) Tools for chemical synthesis in microsystems. Lab Chip 14:3206–3212
33. Gérardy R, Emmanuel N, Toupy T, Kassin V-E, Tshibalonza NN, Schmitz M, Monbaliu JCM (2018) Continuous flow organic chemistry: successes and pitfalls at the interface with current societal challenges. Eur J Org Chem 2301–2351
34. Vaddula BR, Gonzalez MA (2013) Flow chemistry for designing sustainable chemical synthesis. Chim Oggi 31:16–20
35. MacQuarrie DJ (2010) Heterogenized homogenous catalysts for fine chemicals production: materials and processes, Barbaro P, Liguori F (eds). Springer, Dordrecht
36. Granda JM, Donina L, Dragone V, Long D-L, Cronin L (2018) Controlling an organic synthesis robot with machine learning to search for new reactivity. Nature 559:377–381
37. Vámosi P, Matsuo K, Masuda T, Sato K, Narumi T, Takeda K, Mase N (2019) Rapid optimization of reaction conditions based on comprehensive reaction analysis using a continuous flow microwave reactor. Chem Rec 19:77–84. https://doi.org/10.1002/tcr.201800048
38. Wegner J, Ceylan S, Kirschning A (2012) Flow chemistry—a key enabling technology for (multistep) organic synthesis. Adv Synth Catal 354:17–57
39. Carter CF, Lange H, Ley SV, Baxendale IR, Wittkamp B, Goode JG, Gaunt NL (2010) ReactIR flow cell: a new analytical tool for continuous flow chemical processing. Org Process Res Dev 14:393–404
40. Hartman RL, Naber JR, Zaborenko N, Buchwald SL, Jensen KF (2010) Overcoming the challenges of solid bridging and constriction during Pd-catalyzed C-N bond formation in microreactors. Org Process Res Dev 14:1347–1357
41. Kappe CO (2004) Controlled microwave heating in modern organic synthesis. Angew Chem Int Ed 43:6250–6284
42. Leonelli C (2017) Microwave chemistry, Cravotto G, Carnaroglio D (eds). Walter de Gruyter GmbH & Co. KG, pp 39–45
43. Cablewski T, Faux AF, Strauss CR (1994) Development and application of a continuous microwave reactor for organic synthesis. J Org Chem 59:3408–3412
44. Singh BK, Kaval N, Tomar S, der Eycken EV, Parmar VS (2008) Transition metal-catalyzed carbon-carbon bond formation Suzuki, Heck, and Sonogashira reactions using microwave and microtechnology. Org Process Res Dev 12:468–474
45. Baxendale IR, Hornung C, Ley SV, Molina JDMM, Wilkström A (2013) Flow microwave technology and microreactors in synthesis. Aust J Chem 66:131–144
46. Strauss CR (1990) A continuous microwave reactor for laboratory-scale synthesis. Chem Aust 186
47. Chen S-T, Chiou S-H, Wang K-T (1990) Preparative scale organic synthesis using a kitchen microwave oven. J Chem Soc Chem Commun 807–809

48. Kremsner JM, Alexander S, Kappe CO (2006) The scale-up of microwave-assisted organic synthesis. Top Curr Chem 266:233–278
49. Baxendale IR, Hayward JJ, Ley SV (2007) Microwave reactions under continuous flow conditions. Comb Chem High Throughput Screening 10:802–836
50. Glasnov TN, Kappe CO (2007) Microwave-assisted synthesis under continuous-flow conditions. Macromol Rapid Commun 28:395–410
51. Moseley JD, Lenden P, Lockwood M, Ruda K, Sherlock J-P, Thomson AD, Gilday JP (2008) A comparison of commercial microwave reactors for scale-up within process chemistry. Org Process Res Dev 12:30–40
52. Bowman MD, Holcomb JL, Kormos CM, Leadbeater NE, Williams VA (2008) Approaches for scale-up of microwave-promoted reactions. Org Process Res Dev 12:41–57
53. Strauss CR (2009) On scale up of organic reactions in closed vessel microwave systems. Org Process Res Dev 13:915–923
54. Estel L, Poux M, Benamara N, Polaert I (2017) Continuous flow-microwave reactor: where are we? Chem Eng Process 113:56–64
55. Chen S-T, Chiou S-H, Wang K-T (1991) Enhancement of chemical reactions by microwave irradiation. J Chin Chem Soc 38:85–91
56. Raner KD, Strauss CR, Trainor RW, Thorn JS (1995) A new microwave reactor for batchwise organic synthesis. J Org Chem 60:2456–2460
57. Horikoshi S, Abe H, Torigoe K, Abe M, Serpone N (2010) Access to small size distributions of nanoparticles by microwave-assisted synthesis. Formation of Ag nanoparticles in aqueous carboxymethylcellulose solutions in batch and continuous-flow reactors. Nanoscale 2:1441–1447
58. Dallinger D, Lehmann H, Moseley JD, Stadler A, Kappe CO (2011) Scale-up of microwave-assisted reactions in a multimode bench-top reactor. Org Process Res Dev 15:841–854
59. Shieh W-C, Dell S, Repič O (2001) 1,8-Diazabicyclo[5.4.0]undec-7-ene (DBU) and microwave-accelerated green chemistry in methylation of phenols, indoles, and benzimidazoles with dimethyl carbonate. Org Lett 3:4279–4281
60. Shieh W-C, Dell S, Repič O (2002) Large scale microwave-accelerated esterification of carboxylic acids with dimethyl carbonate. Tetrahedron Lett 43:5607–5609
61. Shieh W-C, Lozanov M, Repič O (2003) Accelerated benzylation reaction utilizing dibenzyl carbonate as an alkylating reagent. Tetrahedron Lett 44:6943–6945
62. Savin KA, Robertson M, Gernert D, Green S, Hembre EJ, Bishop J (2003) A study of the synthesis of triazoles using microwave irradiation. Mol Divers 7:171–174
63. Wilson NS, Sarko CR, Roth GP (2004) Development and applications of a practical continuous flow microwave cell. Org Process Res Dev 8:535–538
64. He P, Haswell SJ, Fletcher PDI (2004) Microwave-assisted Suzuki reactions in a continuous flow capillary reactor. Appl Catal A 274:111–114
65. He P, Haswell SJ, Fletcher PDI (2005) Efficiency, monitoring and control of microwave heating within a continuous flow capillary reactor. Sens Actuators B 105:516–520
66. Bagley MC, Lenkins RL, Lubinu MC, Mason C, Wood R (2005) A simple continuous flow microwave reactor. J Org Chem 70:7003–7006
67. Saaby S, Baxendale IR, Ley SV (2005) Non-metal-catalysed intramolecular alkyne cyclotrimerization reactions promoted by focused microwave heating in batch and flow modes. Org Biomol Chem 3:3365–3368
68. Comer E, Organ MG (2005) A microreactor for microwave-assisted capillary (continuous flow) organic synthesis. J Am Chem Soc 127:8160–8167
69. Comer E, Organ MG (2005) A microcapillary system for simultaneous, parallel microwave-assisted synthesis. Chem Eur J 11:7223–7227
70. Shore G, Morin S, Organ MG (2006) Catalysis in capillaries by Pd thin films using microwave-assisted continuous-flow organic synthesis (MACOS). Angew Chem Int Ed 45:2761–27661
71. Baxendale IR, Griffiths-Jones CM, Ley SV, Tranmer GK (2006) Microwave-assisted Suzuki coupling reactions with an encapsulated palladium catalyst for batch and continuous-flow transformations. Chem Eur J 12:4407–4416

72. Glasnov TN, Vugts DJ, Koningstein MM, Desai B, Fabian WMF, Orru RVA, Kappe CO (2006) Microwave-assisted Dimroth rearrangement of thiazines to dihydropyrimidinethiones: synthetic and mechanistic aspects. QSAR Comb Sci 26:509–518
73. Bremnar WS, Organ MG (2007) Multicomponent reactions to form heterocycles by microwave-assisted continuous flow organic synthesis. J Comb Chem 9:14–16
74. Smith CJ, Iglesias-Sigüenza FJ, Baxendale IR, Ley SV (2007) Flow and batch mode focused microwave synthesis of 5-amino-4-cyanopyrazoles and their further conversion to 4-aminopyrazolopyrimidines. Org Biomol Chem 5:2758–2761
75. Shore G, Morin S, Mallik D, Organ MG (2008) Pd PEPPSI-IPr-mediated reactions in metal-coated capillaries under MACOS: the synthesis of indoles by sequential aryl amination/Heck coupling. Chem Eur J 14:1351–1356
76. Bergamelli F, Iannelli M, Marafie JA, Moseley JD (2010) A commercial continuous flow microwave reactor evaluated for scale-up. Org Process Res Dev 14:926–930
77. Dressen MHCL, van de Kruijs BHP, Meuldijk J, Vekemans JAJM, Hulshof LA (2009) From batch to flow processing: racemization of *N*-acetylamino acids under microwave heating. Org Process Res Dev 13:888–895
78. Bagley MC, Fusillo V, Jenkins RL, Lubinu MC, Mason C (2010) Continuous flow processing from microreactors to mesoscale: the Bohlmann-Rahtz cyclodehydration reaction. Org Biomol Chem 8:2245–2251
79. Sauks JM, Mallik D, Lawryshyn Y, Bender T, Organ M (2014) A continuous-flow microwave reactor for conducting high-temperature and high-pressure chemical reactions. Org Process Res Dev 18:1310–1314
80. Öhrngren P, Fardost A, Russo F, Schanche J-S, Fagrell M, Larhed M (2012) Evaluation of a nonresonant microwave applicator for continuous-flow chemistry applications. Org Process Res Dev 16:1053–1063
81. Fardost A, Russo F, Larhed M (2012) A non-resonant microwave applicator fully dedicated to continuous flow chemistry. Chim Oggi 30:14–17
82. Engen K, Sävmarker J, Rosenström U, Wannberg J, Lundbäck T, Jenmalm-Jensen A, Larhed M (2014) Microwave heated flow synthesis of spiro-oxindole dihydroquinazolinone based IRAP inhibitors. Org Process Res Dev 18:1582–1588
83. Kumpiņa I, Isaksson R, Sävmarker J, Wannberg J, Larhed M (2016) Microwave promoted transcarbamylation reaction of sulfonylcarbamates under continuous-flow conditions. Org Process Res Dev 20:440–445
84. Skillinghaug B, Rydfjord J, Sävmarker J, Larhed M (2016) Microwave heated continuous flow palladium(II)-catalyzed desulfitative synthesis of aryl ketones. Org Process Res Dev 20:2005–2011
85. Nishioka M, Miyakawa M, Kataoka H, Koda H, Sato K, Suzuki TM (2011) Continuous synthesis of monodispersed silver nanoparticles using a homogeneous heating microwave reactor system. Nanoscale 3:2621–2626
86. Nishioka M, Miyakawa M, Daino Y, Kataoka H, Koda H, Sato K, Suzuki TM (2011) Facile and continuous synthesis of Ag@SiO_2 core-shell nanoparticles by a flow reactor system assisted with homogeneous microwave heating. Chem Lett 40:1204–1206
87. Nishioka M, Miyakawa M, Daino Y, Kataoka H, Koda H, Sato K, Suzuki TM (2011) Rapid and continuous polyol process for platinum nanoparticle synthesis using a single-mode microwave flow reactor. Chem Lett 40:1327–1329
88. Nishioka M, Miyakawa M, Daino Y, Kataoka H, Koda H, Sato K, Suzuki TM (2013) Single-mode microwave reactor used for continuous flow reactions under elevated pressure. Ind Eng Chem Res 52:4683–4687
89. Matsuzawa M, Togashi S, Hasebe S (2012) Isothermal reactor for continuous flow microwave-assisted chemical reaction. J Therm Sci Technol 7:58–74
90. Yokozawa S, Ohneda N, Muramatsu K, Okamoto T, Odajima H, Ikawa T, Sugiyama J, Fujita M, Sawairi T, Egami H, Hamashima Y, Egi M, Akai S (2015) Development of a highly efficient single-mode microwave applicator with a resonant cavity and its application to continuous flow syntheses. RSC Adv 5:10204–10210

91. Musio B, Mariani F, Śliwiński EP, Kabeshov MA, Odajima H, Ley SV (2016) Combination of enabling technologies to improve and describe the stereoselectivity of Wolff-Staudinger cascade reaction. Synthesis 48:3515–3526
92. Barham JP, Tanaka S, Koyama E, Ohneda N, Okamoto T, Odajima H, Sugiyama J, Norikane Y (2018) Selective, scalable synthesis of C_{60}-fullerene/indene monoadducts using a microwave flow applicator. J Org Chem 83:4348–4354
93. Ichikawa T, Mizuno M, Ueda S, Ohneda N, Odajima H, Sawama Y, Monguchi Y, Sajiki H (2018) A practical method for heterogeneously-catalyzed Mizoroki-heck reaction: flow system with adjustment of microwave resonance as an energy source. Tetrahedron Lett 74:1801–1816
94. Egami H, Sawairi T, Tamaoki S, Ohneda N, Okamoto T, Odajima H, Hamashima Y (2018) (*E*)-3-[4-(Pent-4-en-1-yloxy)phenyl]acrylicc acid. Molbank, M996
95. Yadav VS, Sahu DK, Singh Y, Kumar M, Dhubkarya DC (2010) Frequency and temperature dependence of dielectric properties of pure poly vinylidene fluoride (PVDF) thin films. AIP Conf Proc 1285:267–278
96. The limited examples of microwave flow reactions using non-polar solvents (defined by those with a permittivity $\varepsilon < 5$) use polar/ionic additives, high substrate concentrations or SiC reactor tubes (known to absorb microwave irradiation efficiently; thus acting as a conventional heater) to circumvent the poor microwave absorption of hydrocarbon solvents. See also Refs. 19–22, 62, 74, 75, 77, 79
97. Saida H, Odajima H, Ohneda N, Yokozawa S (2012). SAIDA FDS Inc., World Patent No. WO/2012/043753
98. Barham JP, Koyama E, Norikane Y, Ohneda N, Yoshimura T (2019) Microwave flow: a perspective on reactor and microwave configurations and the emergence of tunable single-mode heating toward large-scale applications. Chem Rec 18:183–203. https://doi.org/10.1002/tcr.201800104
99. Xu D-Q, Yang W-L, Luo S-P, Wang B-T, Wu J, Xu Z-Y (2007) Fischer indole synthesis in Brønsted acidic ionic liquids: a green, mild, and regiospecific reaction system. Eur J Org Chem 6:1007–1012
100. Fitzpatrick JT, Hiser RD (1957) Noncatalytic Fischer indole synthesis. J Org Chem 22:1703–1704
101. An J, Bagnell L, Cablewski T, Strauss CR, Trainor RW (1997) Applications of high-temperature aqueous media for synthetic organic reactions. J Org Chem 62:2505–2511
102. Dubhashe YR, Sawant VM, Gaikar VG (2018) Process intensification of continuous flow synthesis of tryptophol. Ind Eng Chem Res 57:2787–2796
103. Shore G, Organ MG (2008) Diels-Alder cycloadditions by microwave-assisted, continuous flow organic synthesis (MACOS): the role of metal films in the flow tube. Chem Commun 838–840
104. Leadbeater NE, Pillsbury SJ, Shanahan E, Williams VA (2005) An assessment of the technique of simultaneous cooling in conjunction with microwave heating for organic synthesis. Tetrahedron 61:3565–3585
105. Deadman BJ, Collins SG, Maguire AR (2015) Taming hazardous chemistry in flow: the continuous processing of diazo and diazonium compounds. Chem Eur J 21:2298–2308
106. Müller STR, Wirth T (2015) Diazo compounds in continuous-flow technology. Chemsuschem 8:245–250
107. Fuse S, Otake Y, Nakamura H (2017) Integrated micro-flow synthesis based on photochemical Wolff rearrangement. Eur J Org Chem 44:6466–6473
108. Puplovskis A, Kacens J, Neilands O (1997) New route for [60] fullerene functionalization in [4+2] cycloaddition reaction using indene. Tetrahedron Lett 38:285–288
109. He Y, Chen H-Y, Hou J, Li Y (2010) Indene-C_{60} bisadduct: a new acceptor for high-performance polymer solar cells. J Am Chem Soc 132:1377–1382
110. Campisciano V, Riela S, Not R, Gruttadauria M, Giacalone F (2014) Efficient microwave-mediated synthesis of fullerene acceptors for organic photovoltaics. RSC Adv 4:63200–63207

111. Seyler H, Wong WWH, Jones DJ, Holmes AB (2011) Continuous flow synthesis of fullerene derivatives. J Org Chem 76:3551–3556
112. Koyama E, Ito N, Sugiyama J, Barham JP, Norikane Y, Azumi R, Ohneda N, Ohno Y, Yoshimura T, Odajima H, Okamoto T (2018) A continuous-flow resonator-type microwave reactor for high-efficiency organic synthesis and Claisen rearrangement as a model reaction. J Flow Chem 8:147–156
113. Egami H, Tamaoki S, Abe M, Ohneda N, Yoshimura T, Okamoto T, Odajima H, Mase N, Takeda K, Hamashima Y (2018) Scalable microwave-assisted Johnson-Claisen rearrangement with a continuous flow microwave system. Org Process Res Dev 22:1029–1033
114. Harada Y, Sakajiri K, Kuwahara H, Kang S, Watanabe J, Tokita M (2015) Cholesteric films exhibiting expanded or split reflection bands prepared by atmospheric photopolymerisation of diacrylic nematic monomer doped with a photoresponsive chiral dopant. J Mater Chem C 3:3790–3795
115. Kappe CO, Pieber B, Dallinger D (2013) Microwave effects in organic synthesis: myth or reality? Angew Chem Int Ed 52:1088–1094
116. Barham JP, Tamaoki S, Egami H, Ohneda N, Okamoto T, Odajima H, Hamashima Y (2018) C-alkylation of N-alkylamides with styrenes in air and scale-up using a microwave flow reactor. Org Biomol Chem 16:7568–7573
117. The microwave versus thermal comparison result and corresponding permittivity measurements disclosed herein have not been previously disclosed elsewhere
118. de la Hoz A, Díaz-Ortiz A, Moreno A (2004) Microwaves in organic synthesis. Thermal and non-thermal microwave effects. Chem Soc Rev 34:164–178
119. Horikoshi S, Nakamura T, Kawaguchi M, Serpone N (2015) Enzymatic proteolysis of peptide bonds by a metallo-endoproteinaseunder precise temperature control with 5.8-GHz microwave radiation. J Mol Catal B Enzym 116:52–59
120. Tashima S, Nushiro K, Saito K, Yamada T (2016) Microwave specific effect on catalytic atropo-enantioselective ring-opening reaction of biaryl lactones. Bulletin Chem Soc Japan 89:833–835
121. Ichikawa T, Netsu M, Mizuno M, Mizusaki T, Takagi Y, Sawama Y, Monguchi Y, Sajiki H (2017) Development of a unique heterogeneous palladium catalyst for the Suzuki-Miyaura reaction using (hetero)aryl chlorides and chemoselective hydrogenation. Adv Synth Catal 359:2269–2279
122. Weissman SA, Anderson NG (2015) Design of experiments (DoE) and process optimization. A review of recent publications. Org Process Res Dev 19:1605–1633

Chapter 5
Curing of Adhesives and Resins with Microwaves

Robert L. Hubbard and Robert J. Schauer

Abstract The use of microwave energy to heat and cure adhesives, resins, and other polymer precursors is becoming a disruptive and uniquely useful technology. Restrictions in chemical processing are being modified or removed with benefits in product properties and assembly reliabilities. To understand the source of these advantages, it is helpful to review the nature of the inductive polarization effect of electromagnetic energy on electron distribution in molecular orbitals; the kinetic effects of induced bond rotation; the thermodynamic contributions of entropy; and the unique contributions of the microwave wavelength. All of these factors contribute to fundamental changes in the nature of adhesive and resin polymerizations, as well as the chemical, thermal, and mechanical properties of the end products. For a good review of fundamental microwave theory, the 1983 book ***Industrial Microwave Heating*** by Metaxis and Meredith is recommended [1].

5.1 Molecular Polarization and Rotation

Chemical reaction kinetics is dominated by the enthalpy factor, which concerns the exchange of electrons to break and form molecular bonds. The formation of transition states and the energies of activation play the primary role in the polymerization of resins that produces adhesion and structure. The special case of UV activation in radical polymerization will be considered at a later stage.

It is important to consider that molecular dipoles in covalent bonds are the result of differences in electronegativity in atomic electron distributions (Fig. 5.1) [2]. These are "permanent dipoles" as long as the bonds are not broken through a chemical reaction. While the electron attraction forces cause an increase of electronegativity to the right and top of the periodic table, the induced polarizability of electrons in an electromagnetic (EM) field depends on the spacial volume of the bonding electron orbitals. Atomic polarizability increases to the left of the periodic table ($C > N > O > F$) and molecular polarizabilities follow ($CH_4 > NH_3 > H_2O$) as shown in Table 5.1 [2].

R. L. Hubbard · R. J. Schauer (✉)
Lambda Technologies Inc., Morrisville, NC 27560 , USA
e-mail: bschauer@microcure.com

S. Horikoshi and N. Serpone (eds.), *RF Power Semiconductor Generator Application in Heating and Energy Utilization*, https://doi.org/10.1007/978-981-15-3548-2_5

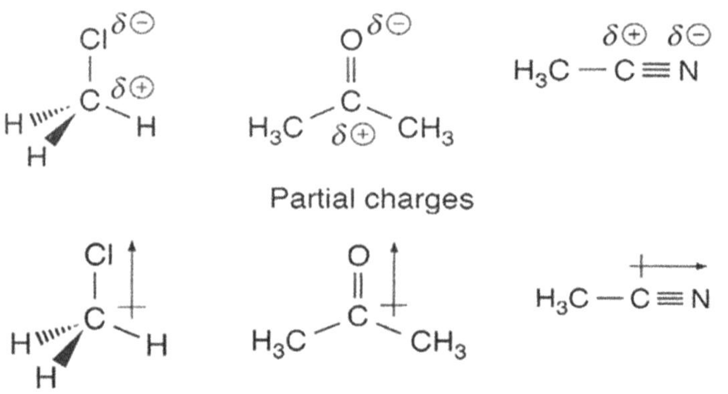

Partial charges

Fig. 5.1 Permanent dipoles in chemical bonding. Reproduced from Ref. [3]. Copyright 2006 by University Science Books

The field effect of induced polarizability is in addition to electronegativity since tightly held electrons are not polarizable. Polarizability also increases down the periodic table (S > O, P > N and H_2S > H_2O). Even though iodine has about the same electronegativity as carbon, the C–I bond is very polarizable and very reactive even without an EM field. It is important to note (see Table 5.1) that hydrocarbons are some of the most polarizable molecules. This is seen in the progressively higher polarizability of molecular bonds of sp^3 carbons in alkanes over sp^2 and *sp* orbitals in alkenes and alkynes, respectively, as well as quaternary aromatic bonding orbitals.

The use of microwave energy is a relatively new method of inducing bond polarizations and of adding an entropy factor that significantly enhances reaction rates, formation of products, and the thermodynamics. Microwave energy interacts with electrons in the highest bonding orbitals of atoms and molecules; however, it is only energetic enough to perturb the distribution of those orbitals, transferring this energy to rotational motion in polarizable bonds [1]. A common example of this polarization preference is found in the water molecule, which has a dipole moment of 1.86 Debye and a polarizability radius of 1.46 Å. This dipolar rotation is often quoted as the primary heating mechanism of microwave energy; it should be noted, however, that the symmetrical methane molecule with a dipole moment of 0 Debye has an even larger polarization radius of 2.6 Å because of the more extended sp^3-hybrid C–H bond lengths. This accounts for the less energetic but significant rotation of the backbone and side chains, even in nonpolar polymers. This relatively low-energy redistribution with microwave energy is still much too low to cause the breakage of covalent bonds between atoms in molecules (decomposition).

Table 5.1 Atomic and molecular polarizabilities ($\alpha/10^{-24}\ cm^3$)

Atomic polarizabilities

H	0.6668									He	0.205
		C	1.76	N	1.1.	O	0.802	F	0.557		
				P	3.13	S	2.90	Cl	2.18		
								Br	3.05		
								I	4.7 (or 5.35)		

Selected molecular polarizabilities

CH_4	2.6	NH_3	2.21	H_2O	1.45	H_2S	3.8
CO_2	2.91	CS_2	8.8	CF_4	3.84	CCl_4	11.2
C_2H_2	3.6	C_2H_4	4.25	C_2H_6	4.45	CH_3OH	3.23
Benzene	10.32	Cyclohexene	10.7	Cyclohexane	11.0		

Reproduced from Ref. [2]. Copyright 1990 by CRC Press Inc.

5.2 Reaction Kinetics

The well-known fast reaction kinetics resulting from the use of microwave ovens can be understood from this rotational energy applied in the collision form of the Arrhenius equation as follows (Eq. 5.1):

$$k = Z * \rho * \exp\left(\frac{-E_a}{RT}\right) \quad (5.1)$$

where k is the reaction rate, Z is the collision frequency, ρ is the steric factor, E_a is the energy of activation, R is the gas constant, and T is the Kelvin temperature. With enhancement of rotational energy from microwave energy, the collision frequency is higher and the statistical probability of optimal arrangement of reacting chemical orbitals (steric factor) is higher. An argument could be made that the increased reaction kinetics is actually due to decreased energy of activation, but this is less convincing in the condensed phases that will be discussed here [4]. At the same reaction temperature, it has been shown that equivalent chemical reactions can be run at about ten times the rate of standard thermal reactions [5].

Additionally, with the increase of the pre-exponential factors in Eq. 5.1, it should be obvious that the evaporation of water or organic solvents can be performed at lower temperatures. The solvent *n*-methylpyrolidone (NMP) boils at 140 °C rather than the usual 203 °C with the application of microwave energy. This increased dispersion of solvents makes the drying of materials much more efficient with microwave energy.

An apparently kinetic-controlled reaction where the favored reaction products were quite different from microwave heating than with standard thermal heating has been described [5] although the authors misinterpreted the results [6]. Figure 5.2 portrays the competitive reaction that, with a convection oven heating, produced an even mixture of both an amide (top) from an addition reaction and an ester (bottom) from an insertion process [5]. The microwave-heated reaction actually produced exclusively the ester product. Since the amide product can rearrange to produce the ester product, it is more likely that there is a lower-energy transition state for the rearrangement so that the microwave reaction produced the ester exclusively by kinetic control. Although the reaction was not investigated further, it is possible that a lower-energy (lower temperature) microwave-assisted reaction may produce a mixture or, even exclusively, the amide.

5.3 Thermodynamics

The Gibbs free energy change (ΔG°) of a chemical reaction (Eq. 5.2) includes both an enthalpy term (ΔH°) and an entropy term (ΔS°). The enthalpy difference between starting materials and products is of primary interest in determining how much bond energy is added to a system or produced in a chemical reaction under steady state

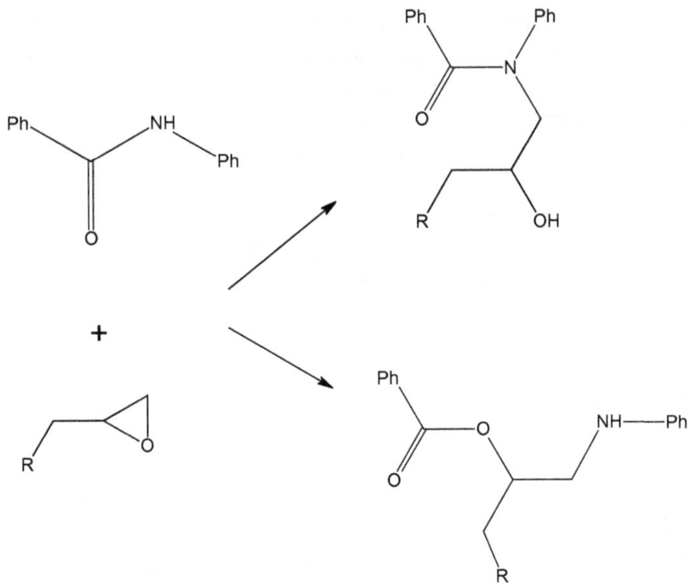

Fig. 5.2 Amide–epoxide addition/insertion reaction. Reproduced from Ref. [5]

conditions. The entropy is usually a small term and is often ignored unless the reaction product has either formed a ring or opened a ring thus increasing the total bond degrees of freedom.

$$\Delta G^\circ = \Delta H^\circ - T\Delta S^\circ \tag{5.2}$$

Since the enthalpy is related to the energy of activation in Eq. 5.1, it could be argued that the rotational energy added with microwaves should contribute to the Gibbs free energy as additional entropy. A thermodynamic-controlled reaction should also increase the activation of a chemical system. In fact, it has been determined that chemical reactions conducted with microwave energy can be completed at 40% lower temperatures or more [7].

5.4 Field Size and Uniformity Effects

Like any single-frequency electromagnetic field (or sound waves) injected into a cavity with parallel walls, resonant nodes of high energy and low energy will be established quickly. Not only does this produce uneven heating patterns, it also causes high-energy arcing on any metallic surface present. It is clear from Eq. 5.3 that since the average power dissipation (P_{av}) is directly dependent on the square of the applied field intensity (E^2_{rms}), the non-uniformity of power intensity at nodes is even worse. Early published experimental work with microwave fields was difficult to conduct, and the results were highly variable [8–10]. Reaction cells were designed to be small (1 mm^3) to approximate one node; however, the resulting cures were still very non-uniform as evidenced by thermomechanical analyses. It was necessary to develop a means for producing a large and uniform microwave field for practical applications.

$$P_{av} = \omega \varepsilon_0 \varepsilon''_{eff} E^2_{rms} V \tag{5.3}$$

5.5 Variable Frequency Microwaves {VFM}

The variable frequency microwave technology was developed at Oak Ridge National Laboratories in 1991 [11] and put into volume production use in 2002 by Lambda Technologies. A large uniform field is produced by changing the injected field frequency 4096 times each 0.1 s and then reset to start again. This produced a constantly changing node pattern every 25 μs. This rapid cycling pattern is not only a uniform microwave field, but no charge on a metal surface can be built up in such a short time before the node pattern changes. Another feature is that large volume fields of 1 m^3 can be produced in a practical space. At this point, it became possible to conduct experiments into the nature of chemical reactions effected by uniform microwave fields.

Of course as soon as a material is placed into this VFM field, the uniformity of the field and power dissipation is affected as per Eq. 5.3. With only a very small volume of material, the power dissipation will be correspondingly low. For example, the attempt to cure a thin layer of resin on a small metallic object (e.g., a syringe needle) will be impractical. The heat from the energy absorbing on the film will be transferred inductively to the metal and immediately be convectively transferred to the air environment. An unexpected result is that the larger the volume of absorbing material, the higher the efficiency of heat transfer is (Eq. 5.3). In the general sense, the holding fixtures for samples in a VFM field should be chosen with low thermal conductivities (poorly heat-sinking) and with a low loss modulus (E'').

The use of a rapidly scanning frequency field also allows the power level to be applied in a closed-loop feedback system with the measured temperature of the target material. In this manner, the power can be adjusted automatically in real time to match the programmed sample temperature. Not only is this a direct method of applying only the needed power to heat the samples to the required temperature, but it also makes the total applied power subject to the total number of samples placed in the field (the volume V in Eq. 5.3). The number of samples in any run does not require manual adjustment of the power.

5.6 Temperature Control of Adhesion

Another property of microwave energy is the longer wavelength vis-à-vis infrared or visible wavelengths that are more commonly used for heating molecules. Infrared energy heats by bending and stretching chemical bonds; however, its wavelength is around 1–2 mm which only causes the penetration depth to be rather shallow. Heating to greater depths relies on induction through the bulk as displayed in the graphic sequence of Fig. 5.3.

Microwaves heat by molecular rotation with a wavelength of tens of millimeters which produces a penetration depth of several meters through organic materials (at the speed of light). All the molecules in a typical sample of an adhesive are rotated/heated at exactly the same instant in all dimensions, as illustrated in the graphic sequence of Fig. 5.4.

Most adhesion mechanisms depend on chemical reactions that produce highly polar functional groups that are attractive to surfaces of metals, glasses, and other organics. The epoxide reaction (Fig. 5.5) [12] is a good example with the creation of one hydroxyl function that bonds with metals, glasses, and organics for every reaction. With the higher reaction kinetics of VFM, bonding functional groups are being produced nearly instantly throughout the volume of the sample.

There are adhesion reactions that occur at all surfaces of the adhesives at the earliest stages of heating. Adhesion is linear with time everywhere: as much as six times more effective [13].

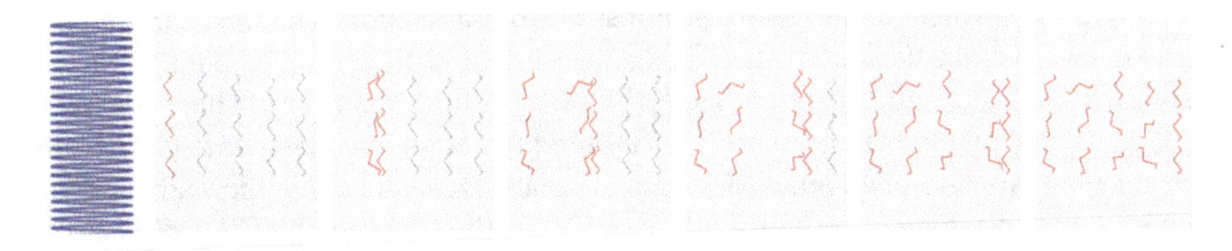

Fig. 5.3 Infrared heating at the surface and heat penetration by induction to the other side. Copyright 2018 by R. L. Hubbard

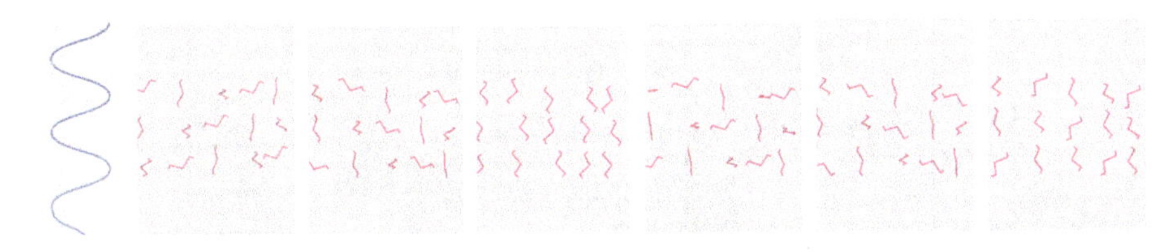

Fig. 5.4 Instantaneous microwave heating by rotation of all molecules simultaneously. Copyright 2018 by R. L. Hubbard

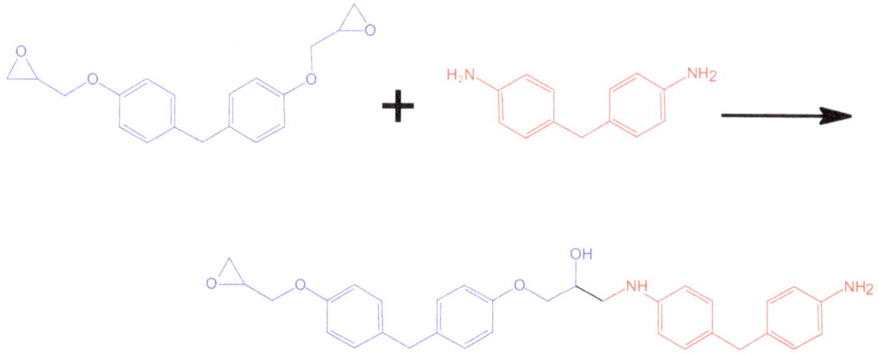

Fig. 5.5 Reaction of an epoxide and amine to form a hydroxyl adduct. Reproduced from Ref. [12]

5.7 Unique Polymerization Characteristics

The extent of polymerization of a polymer precursor proceeds to completion with the completion of the sum possible of all chemical reactions. As polymers become more highly rigid, some of the reactions are blocked by steric hindrance. As a result, very few polymers attain 100% chemically reacted so that a ***full extent*** of polymerization is more typically in the range of 90–98% reacted depending on the type and chemistry of the starting materials. This leaves unreacted functional groups that limit the final thermomechanical, chemical, and electrical properties of the polymer. Additional heating of a polymer can then only progress to decomposition. The most commonly used measure of the "extent of cure" is the measure of its transition with temperature from a glassy state to a rubbery state (glass transition point Tg). There are several analytical methods to measure Tg; (Differential Scanning Calorimetry, Thermo Mechanical Analysis, Dynamic Mechanical Analysis) each having their advantages but each also producing a slightly different Tg for the same sample (Fig. 5.6). As a sample is cured at higher temperatures, the extent of cure increases until the measured

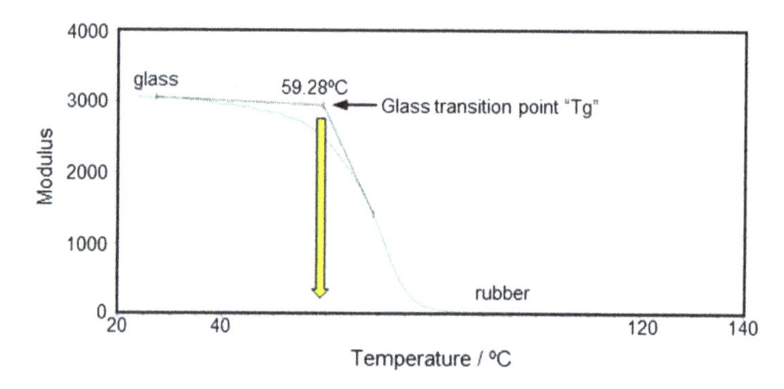

Fig. 5.6 DMA analysis of the glass to rubber transition of a polymer. Copyright 2018 by R. L. Hubbard

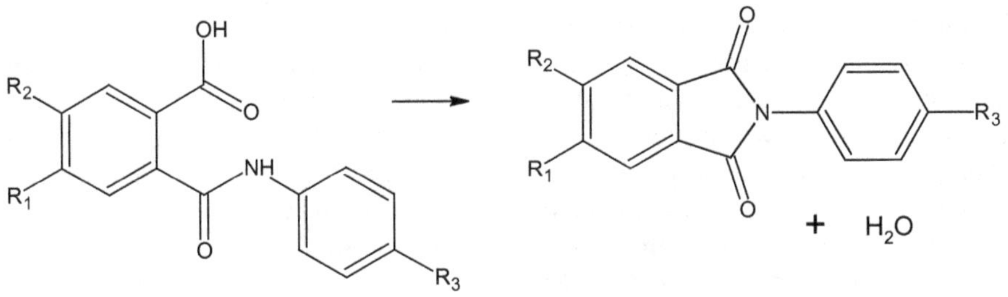

Fig. 5.7 Polyamic acid precursor reaction to form polyimide thermoplastic films. Copyright 2014 by R. L. Hubbard

Tg stops increasing. That point is referred to as the "ultimate Tg" or Tg∞. If the sample is heated further, decomposition begins which decreases the Tg. Adhesives are often listed with a Tg value, however, that does not guarantee that the full extent of cure (Tg∞) is being described nor, in fact, known.

5.7.1 *Thermoplastic Chemistries*

Thermoplastic polymers have the properties of conforming to surfaces that can be re-formed to new shapes after being heated to the rubbery state and back. The long-chain precursors commonly have structures along the chain that react to produce rigidity which increases Tg and provides thermal and chemical stability of films and coatings. An example is the polyimide family which has Tg values at or above 300 °C and is very chemically stable when fully cured (Fig. 5.7). As with all polymers, the cure temperatures must be kept higher than Tg for the polyamic acid or polyamic ester functional groups to be cyclized to polyimides. Cures can approach 100%, although analytical measurements are not accurate above 95%. Standard oven cure processes include temperatures of 350–380 °C and processing times of 4–6 h for the polyimide coatings used on almost all of the microelectronic silicon devices now in use.

5.7.2 *Microwave Curing of Thermoplastics*

As might be expected, microwave curing of polyimides allows for much faster cure times (5 min) [14]. With the large uniform fields of VFM, even the coatings of large silicon wafers can be fully cured across the wafer surface. The polybenzoxazole (PBO) family of polymers is also used as a wafer dielectric layer. They have the advantage of aqueous photolithographic development, and some have lower process temperatures and Tg values less than 300 °C. Both the polyimides and PBOs can be fully cured with microwaves at substantially lower temperatures than their Tg values [15, 16]. The polyimides can be cured at a temperature 150 °C lower than the standard oven cure, while the PBOs can be cured at 50–100 °C lower. A chain extension reaction which increases the Tg values of PBOs can compete with the cyclization reaction. VFM appears to preferentially complete the cyclization reaction of the PBO chemistry before producing the chain extension [17]. As a result, some of these PBO materials can cure as low as 170 °C, especially if the starting materials and conditions are optimized for VFM [16].

5.7.3 Thermoset Chemistries

Thermoset polymers are produced from small starting molecules that are cross-linked to rigid networks that on cooling resort to their original shape after being heated to the rubbery state. There are usually two components with one component having a functionality of at least three or four. Often the resin component is dominant and the higher functionality component is incorrectly termed a catalyst or hardener. As the cure begins, the network has the low viscosity of a Newtonian liquid. Gelation occurs when a chain is produced that reaches across the entire network. This greatly increases the viscosity to that of a Hookean solid (Fig. 5.8). The remaining smaller molecules must disperse through the bulk to connect and cross-link with the network backbone ("infinite polymer"), and this requires ever increasing temperatures. As the cure is completed, the fracture energy also increases, thereby resulting in lower brittleness.

The cure temperature must be increased to always be higher than the incipient Tg of the network being formed or else vitrification will occur [19] as shown in the

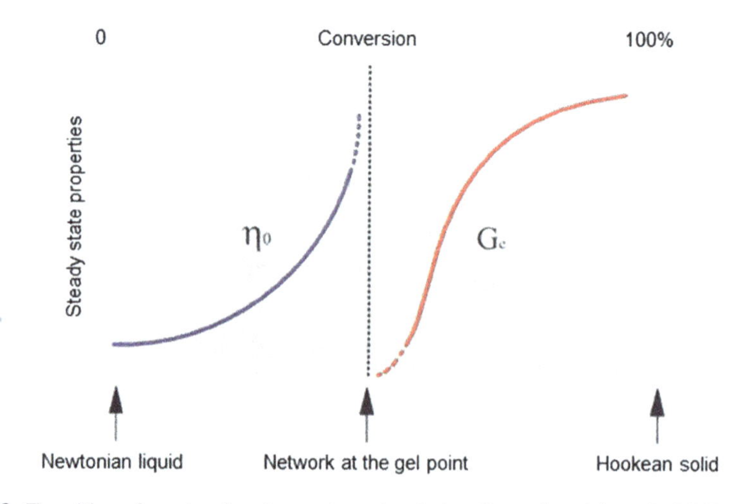

Fig. 5.8 Transition of a network polymer through gelation. Reproduced from Ref. [18]

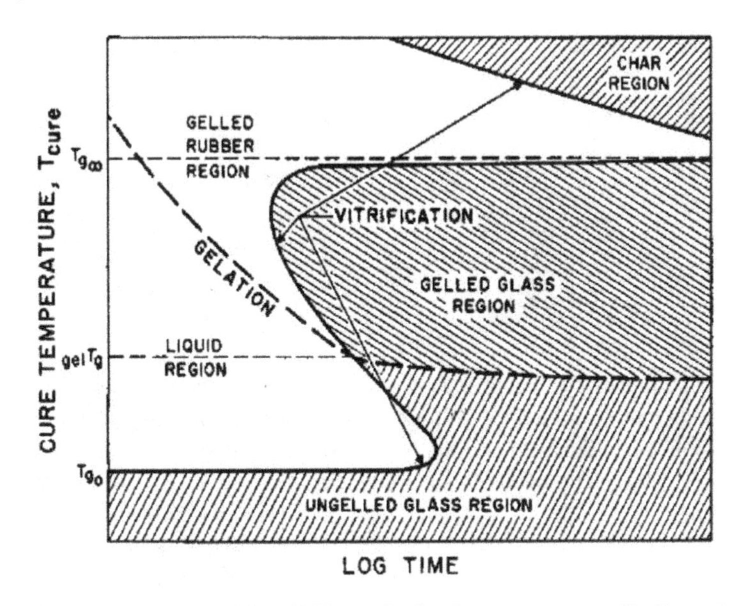

Fig. 5.9 Time temperature transition (TTT) graph of a thermoset cure profile. Reproduced from Ref. [19]

time temperature transition (TTT) diagram of Fig. 5.9. The vitrified, or glassy state, inhibits further cure, regardless of time, and produces an unstable temperature condition. The glass appears to be cured, but if heating above the network Tg occurred, the reactions would continue and the thermal, chemical, and mechanical properties of the glass would evolve until full cure was reached. To avoid the unstable vitrified state, cure temperatures should be chosen to exceed the ultimate Tg∞ by at least 15–20 °C. Since the cure time is a practical issue, the cure temperature of commercial thermosets is often raised to 50 °C above the Tg∞. Note that the x-axis in Fig. 5.9 represents the logarithm of time, which makes the increase of time, rather than the increase of temperature, much less practical.

It is important to note that vitrification has been investigated only recently so that many of those who work with thermoset adhesives and molded composites are unaware of the effects of under-curing on the properties expected from their materials. In an attempt to use lower cure temperatures with longer times, they are producing vitrified glasses that only appear cured; however, they will be chemically, thermally, and mechanically unstable (see discussion below).

5.7.4 Microwave Curing of Thermosets

With the enhanced kinetics of microwave energy, it is possible to speed the cure of thermosets by a factor of 10× or more [20]. The higher the temperature, the

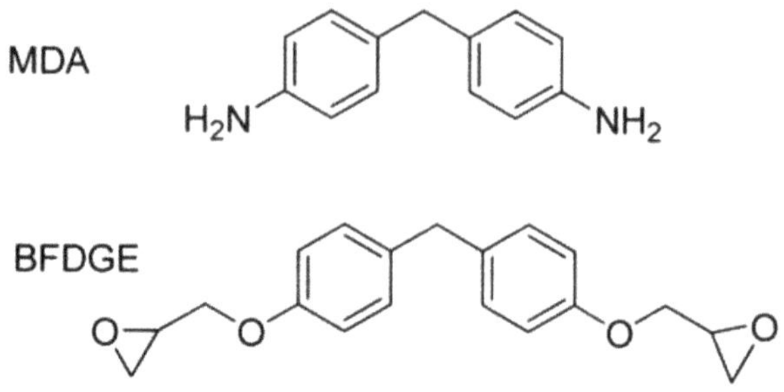

Fig. 5.10 Basic amine and epoxide adhesive components. Reproduced from Ref. [12]

faster the cure rate is but rather than a somewhat exponential increase in time, the microwave cure is more linear with time. This has been a useful method for improving throughput for adhesives and encapsulation in high volume manufacturing for more than 15 years. The capability to reduce the curing temperature of adhesives, encapsulants, and molding compounds by the use of microwaves is a breakthrough in manufacturing flexibilities. The example to be illustrated here is the cure of epoxide-amine chemistry which has been thoroughly studied in the literature and industry for more than fifty years because of its general usefulness. A good review of the oven cure properties of the bisphenol-F diglycidyl ether (BFDGE) and methylene dianiline (MDA) (Fig. 5.10) is given elsewhere [12] including several methods for the measurement of vitrification when cure temperatures less than the $Tg\infty$ of 150 °C were used. It cannot be too strongly stated that the full cure of thermosets cannot be accomplished with conventional methods without cure temperatures above $Tg\infty$. Inadequate curing degrades the thermal, chemical, and mechanical properties of the polymers.

By contrast, VFM cures of these same starting materials were completed with identical properties when they were cured at temperatures as low as 100 °C [7]. Furthermore, it was found that the initial cure reactions before gelation of the oven and VFM cures were quite different. With the oven cure at 170 °C, there was about an even mix of linear chain building and cross-linking between chains as directly determined by quantitative ^{13}C FT-NMR. With VFM cure at 120 °C, there was mostly linear chain building with very little cross-linking. The $Tg\infty$, the modulus, the lack of vitrification, and the cross-link density were the same for all the fully VFM-cured polymers. Unpublished data by the author confirms that the substitution of the common bisphenol-A diglycidyl ether (BADGE) for the BFDGE above produced the same pre-gelation result of preferred linear chain building to cross-linking with VFM compared to oven cure.

5.7.5 *Commercial Adhesive Results*

Commercial proprietary epoxide blends used for flip-chip underfill applications have been shown to cure with VFM at lower temperatures than the recommended oven

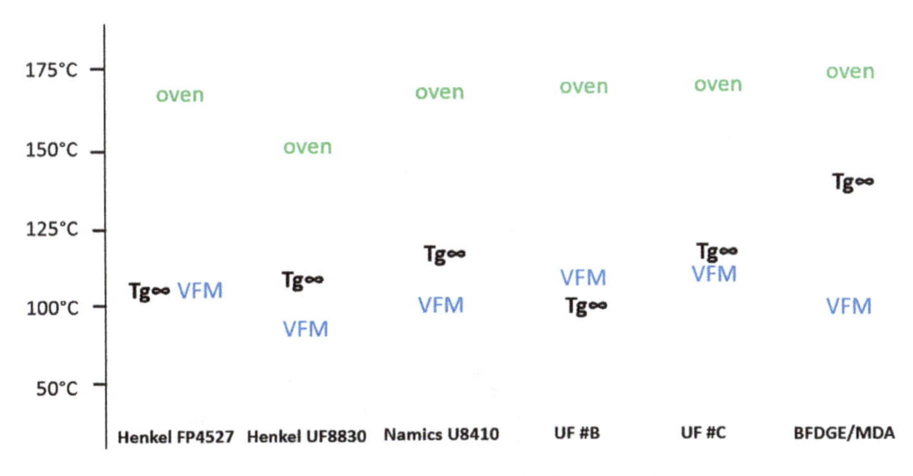

Fig. 5.11 Cure temperatures for commercial epoxide blends with equivalent Tg∞ values. Reproduced from Ref. [20]

cures (Fig. 5.11). These blends were cured to the same indicated Tg∞ at the indicated "oven" and "VFM" cure temperatures [19]. In all cases, the VFM-cured samples required less time even at the lower temperatures. All of the recommended oven cures could have been reduced by extending the cure time (exponentially) but lowering the oven cure temperature below Tg∞ produced vitrification. The two-part epoxide-amine data discussed above is included in the graph for reference.

5.8 Stress and Temperature

The chemical shrinkage stress created during the polymerization reaction as the temperature is lowered from cure temperature to room temperature is also depicted in Fig. 5.12 [21]. Even though there is some stress created in the liquid state before gelation (dashed blue line), the more important stress factor in a thermoset begins at gelation and increases with cooling. Note that viscosity decreases along the ordinate. Not shown is the exponential effect of time on cure and viscosity that was depicted in Fig. 5.9.

If the cure temperature was reduced (Fig. 5.13), the chemical shrinkage stress would be reduced substantially [21]. Note that the cure temperature in both Figs. 5.12 and 5.13 is still higher than the Tg∞ (100 °C in this example at letter *E*). Lowering the cure temperature is therefore a very effective method to reduce chemical stress in adhesives and resins. Unfortunately, few practical applications will choose cure times in excess of an hour or two, much less days. The option does exist, however, to use VFM to reduce cure temperatures substantially, while still having cure times of less than an hour.

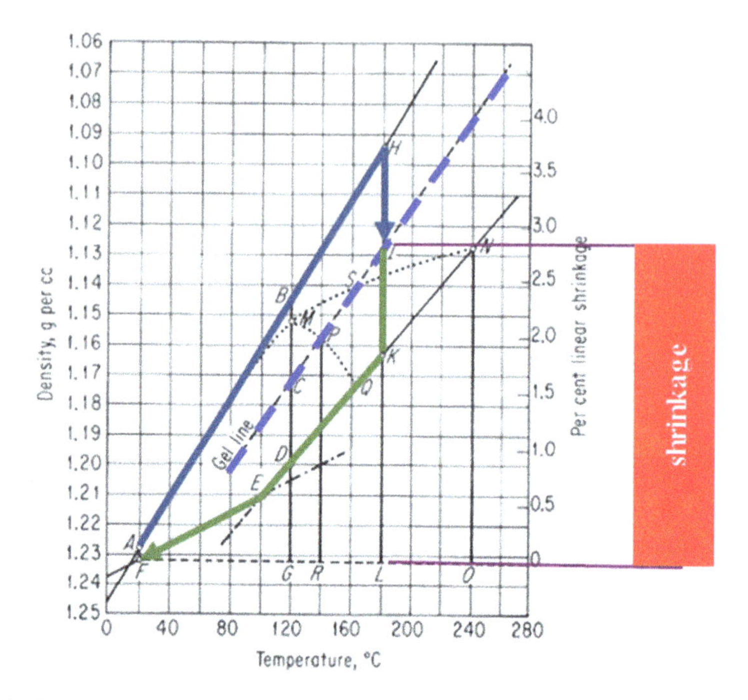

Fig. 5.12 Chemical shrinkage stress begins at gelation temperature. Reproduced from Ref. [21]

Mechanical stress is also a temperature-dependent phenomenon. Whenever two materials are joined that have different coefficients of thermal expansion (CTE), there will be tension stress created at the two interfaces with the adhesive. An example is the bonding of silicon (3 deg/ppm) to an organic substrate (18 deg/ppm) in Fig. 5.14. As the temperature changes from the adhesion cure (165 °C) to room temperature, the organic substrate shrinks a factor of six times more than the silicon. This causes enough compression (2100 ppm) in the silicon to shatter its surface. If the cure temperature was reduced to 100 °C, the tension would then be reduced to 1125 ppm and the silicon would survive the stress if it were not too large. In addition, as the dimensions of the silicon increase, the reliability of the silicon and the joint decreases, especially at the points farthest from the center (neutral point).

The reliability of an IBM large microprocessor package was doubled by using VFM to cure the underfill adhesive at 115 °C rather than 165 °C [23]. An additional benefit with VFM is the elimination of the substrate drying step before the introduction of the underfill dispense and cure steps [24]. The use of low temperature VFM adhesive curing for package-on-package applications is also described in a patent from Taiwan Semiconductor Manufacturing Corporation [25]. The primary benefit was listed as the reduced warpage from CTE mismatch in the two assemblies.

It is clear that the cure temperature of adhesives and resins in practical applications is a very important factor in terms of reliability, but it is also a factor in the cost and time of processing. Lowering process temperatures enables more sensitive

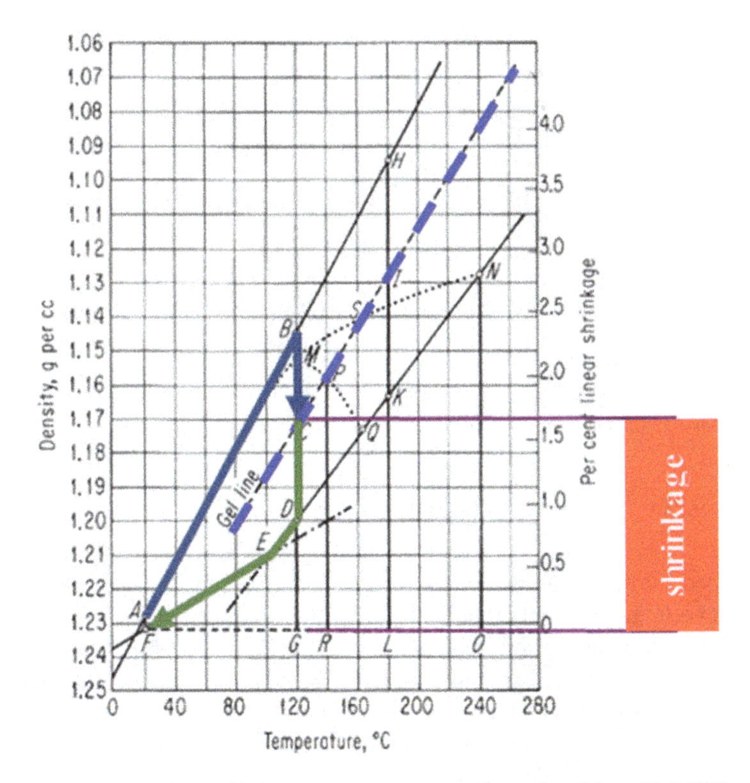

Fig. 5.13 Reduced shrinkage with lower cure temperature. Reproduced from Ref. [21]

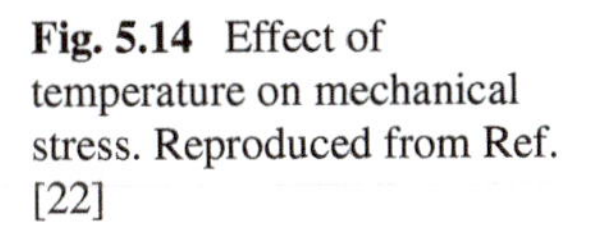
Fig. 5.14 Effect of temperature on mechanical stress. Reproduced from Ref. [22]

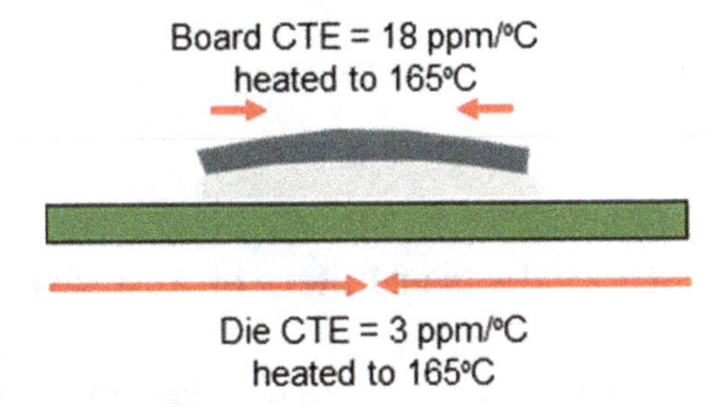

materials and parts to be used and more efficient costs to be realized. Unfortunately, an understanding of the restrictions imposed on cure temperatures is not common. Adhesives, encapsulants, and mold compounds are often listed with Tg values well in excess of the cure temperatures.

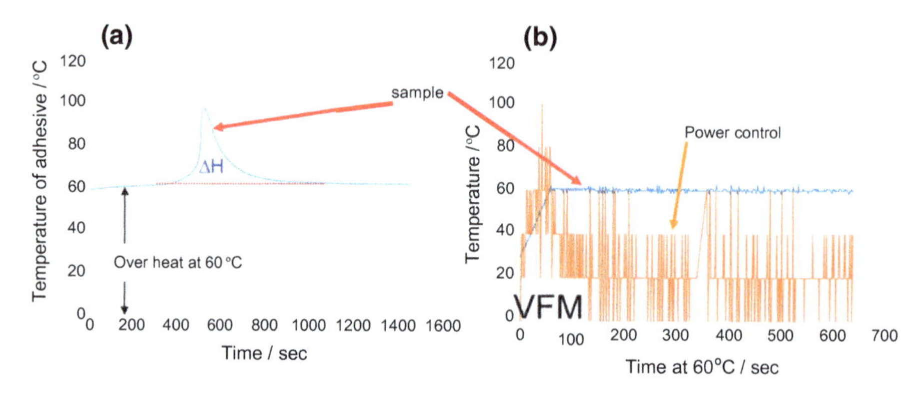

Fig. 5.15 Exothermic effect and its control. Copyright 2018 by R. L. Hubbard

5.9 Exothermic Temperature Control

As described in the thermodynamics section above, a chemical reaction that produces heat (ΔH°) will supply that heat to neighboring molecules in a gradual wave through the bulk when heated in a standard oven (Fig. 5.3). Not observed in most commercial ovens is the effect that ***exothermic*** heat has on the actual temperature of the sample (Fig. 5.15a), which is 97 °C rather than the expected 60 °C from the oven air temperature. This additional temperature adds shrinkage stress to the resin and potential damage to parts that might be temperature sensitive. By contrast, the sample temperature in Fig. 5.15b is controlled in the VFM environment to the programmed 60 °C by the digital closed-loop control of temperature with power in VFM. The heating mechanism by electromagnetic waves that travel at the speed of light allows this controlled temperature to reach all of the sample molecules instantaneously (Fig. 5.4).

5.10 Effects of Fillers in Adhesives

All of the commercial blends in Fig. 5.11 included silica particles, elastomers, and colorant fillers. The addition of micron-sized silica spheres is commonly used to lower the coefficients of thermal expansion (CTE) of the blend, which in turn should better match the adhesive (60 deg/ppm) to the silicon and substrate being joined. This would theoretically lower the stress in the joint from the CTE mismatch of the parts as shown in Fig. 5.14. In Fig. 5.16 [26], no decrease in die warpage results from an increase in silica loading with oven curing (red circles). Higher modulus represents an increase in stress within the material, so suppliers compensate by decreasing the modulus by adding elastomeric fillers. There has also been a trend to lower the modulus of the starting chemistries; however, this results in lower $Tg\infty$ polymer networks. These stress trade-offs directly affect the reliability of electronic

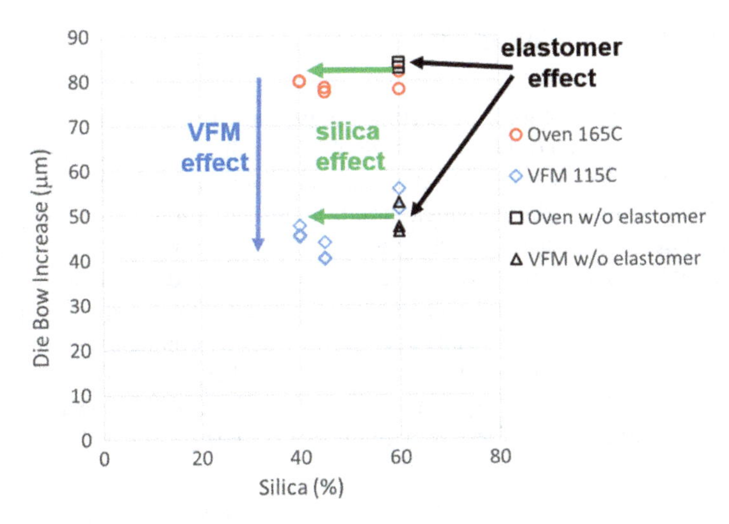

Fig. 5.16 Die bow versus cure temperature and silica loading. Reproduced from Ref. [27]

assemblies. The shift to brittle low-k dielectrics has also been shown to require low modulus adhesives at the expense of the lower values of Tg∞ [27].

The effect of fillers on the VFM curing of a commercial blend is shown in Fig. 5.16. Whether the silica loading is 40 or 60%, it seems to have only a small effect on the measured bow of a 20 mm die on a 35 mm substrate using either oven cure or VFM. The same is true for whether there was a 15% elastomer loading or not in the blend. The only significant reduction in die bow was from the use of VFM curing at 50 °C lower temperature. The storage modulus (Table 5.2) of these samples above Tg∞ is directly proportional to the cross-link density of the blend which suggests that microwave curing increases the molecular chain length between cross-links in the final polymer. This was not the case for the simple BFDGE/MDA or BADGE/MDA mixtures without additives so this effect is likely to have a different mechanism.

Table 5.2 Effect of curing type on modulus and warpage

Cure type	Elastomer	Elastic modulus (MPa)	Wafer bow increase (μm)
Oven	Y	71	79.4
Oven	N	72	83.5
VFM	Y	40	44.1
VFM	N	34	48.9

Reproduced from Ref. [27]

5.11 Application Failure Mechanisms

For any application, the choice of starting components for an adhesive or coating is often made to suit a high or low Tg and suitable modulus. Since these two variables are often confounded, it is usually the case that has a desirably high Tg result in an undesirably high modulus, which results in less plasticity and lower fracture strength. As described earlier, adhesion to surfaces is proportional to the extent and homogeneity of the cure. With oven cures, the efforts to produce adequate adhesion are focused on a secondary cure step at higher temperatures. Vitrification produces glass structures that appear hard and fully cured; however, it includes as many defects as there are incomplete cure sites. It is ironic that what appear to be over-heated polymers with cracks are actually under-cured. Complete curing at low temperatures and reasonable times is now feasible with microwave energy.

5.12 Interactions with UV Adhesives and Microwaves

Adhesives designed to cure by UV light exposure provide a very fast reaction but at a thinner penetration depth of only microns. This provides a cured surface, but the bulk is not completely cured so a second polymer is often blended that is heat cured at higher temperatures and longer times if needed. Microwaves, and specifically VFM, will not initiate the free-radical reaction but can be used to increase the speed or lower the temperature of an included heat-cure polymer. Because of the intermixture of the two resins, the cure of the surface by UV is often only about 30% complete. Even though VFM will complete the secondary resin cure, the resultant non-homogenous polymeric network will not easily be completely cured. The result will have a lower Tg than either of the UV or the heat cured and will have lower fracture strength due to the higher number of defects and fracture pathways. An example of the blended thermomechanical properties is shown in Fig. 5.17.

The initial cure with UV (in blue) produces a partial cure with the overall network having a Tg of 75 °C. After an additional VFM cure at 100 °C (in red), the network Tg is mostly 100 °C with a substantial area having Tg = 120 °C. Even higher temperature curing with VFM (or oven) does not increase the bulk Tg of this sample. This probably depicts two polymer matrices intertwined together without the chance of complete cure. Further study might compare the fracture strength of this two-step approach to completing the cures by using a very fast (seconds) VFM cure of an optimized thermal cure adhesive.

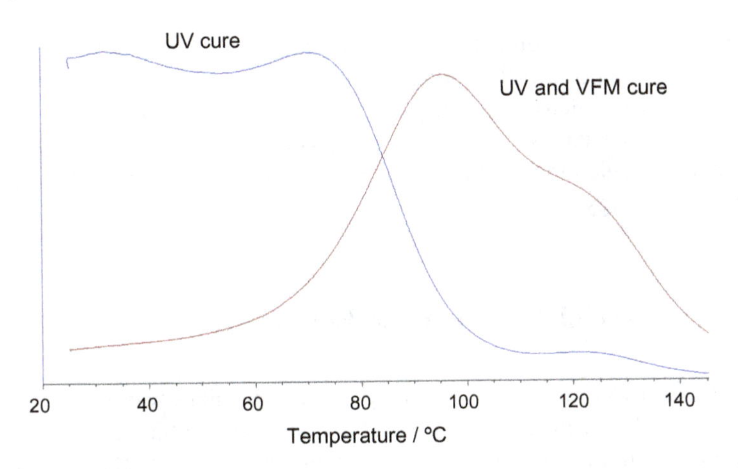

Fig. 5.17 UV cure and UV + VFM cure. Copyright 2018 by R. L. Hubbard

5.13 Chemical Optimization for Microwaves

It has been shown that even the much higher reaction rates of VFM for polymerization can be improved substantially by the choice of more receptive resin designs. In Fig. 5.18, the increased reaction rate of VFM (blue curve) over oven cure (red curve) of the same resin is contrasted with a resin specifically designed for VFM. The designed resin would not cure faster, or at lower temperature, in a standard oven however.

Until recently, industry implementation of VFM has been limited to the curing of available resins already qualified for production use. With the possibility of a substantial lowering of cure temperatures and/or using much faster cure times, there

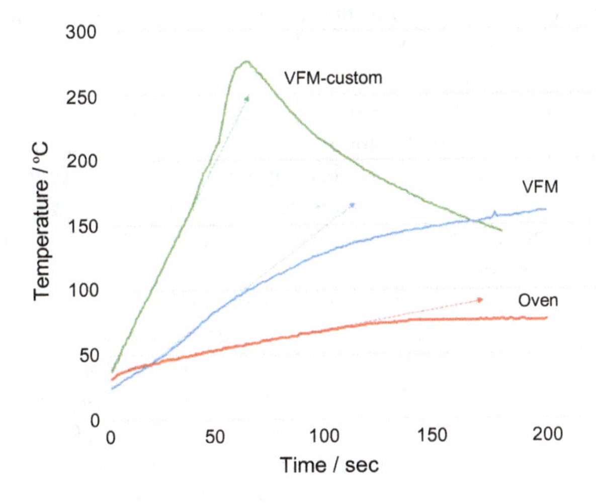

Fig. 5.18 Same resin cured by oven and VFM. Modified resin cured by VFM. Copyright 2018 by R. L. Hubbard

is a strong interest in modifications by resin suppliers. Since the modifications are predictable and do not involve additives, there is a predicted more general interest in the chemistry of microwave reactions. The same experimental methods used by chemists for many decades can be used to design molecules with optimized properties with the added accelerated reaction rates and lowered reaction temperatures provided by microwave energy.

5.14 Commercial Adoption and Equipment

Microwave ovens are used in most homes as a convenient way to quickly reheat food or drinks. The primary benefits for the home user are that the food (or drink) heats while the air, dish, or cup does not heat and that the user gets to eat (or drink) sooner than if the food was heated by conventional means. The idea for consumer microwave heating is that water is common to most drinks and foods that people consume, most of whom are not aware that microwaves also heat meat tissues and other biological polymers, not to mention some of the plastic packaging. Microwave ovens have been used successfully commercially for food preparation, drying grain for agriculture, drying lumber, and for removing water from pharmaceutical powders. Accordingly, this section will address the challenges associated with using microwave ovens commercially in other industries.

With the exception of variable frequency microwave (VFM) technology, originally developed at Oak Ridge National Laboratory, virtually all other commercial microwave ovens operate with fixed frequency microwaves (FFMW). Such FFMW ovens operate at 915 MHz, 2.45 GHz and occasionally at a frequency of 5.8 GHz. These FFMW ovens produce hot and cold spots that pose challenges for commercial adoption. Rotating the target material to be heated and/or incorporating mode stirrers can mitigate some of the inherent non-uniformities in FFMW ovens, thus making it suitable for some, but not all, applications if uniformity was not critical and there were no metallic surfaces that would produce arcing. Only VFM technology has been adopted commercially in the electronics and semiconductor industries. Unlike FFMW, VFM sweeps rapidly through a broad bandwidth of microwave frequencies to electronically stir the hot spots (or modes). Rapid electronic mode stirring with VFM produces uniform heating without the need to move the target materials to be heated and also eliminates conditions that cause arcing to electrically conductive materials. Large volume cavities have also been shown to have good uniformity for large assembly manufacturing.

The MicroCure5130 depicted in Fig. 5.19 is an inline conveyor VFM system used in the high volume production of electronic assemblies. Pallets loaded with parts are automatically loaded into and out of the oven for curing or drying processes.

Under funding from the US Department of Energy, Lambda Technologies developed the VariDry200 system as shown in Fig. 5.20 for continuous in line roll-to-roll drying of wet coatings on metal foil. Benefits are faster drying times, better adhesion, and thicker coatings. The VariWave1200 system shown in Fig. 5.21 is used for labo-

Fig. 5.19 MicroCure5130 inline conveyor VFM oven. Reproduced from Lambda Production folder, Copyright 2017 by R. J. Schauer

Fig. 5.20 VariDry200 continuous roll-to-roll dryer. Reproduced from Lambda Production folder, Copyright 2018 by R. J. Schauer

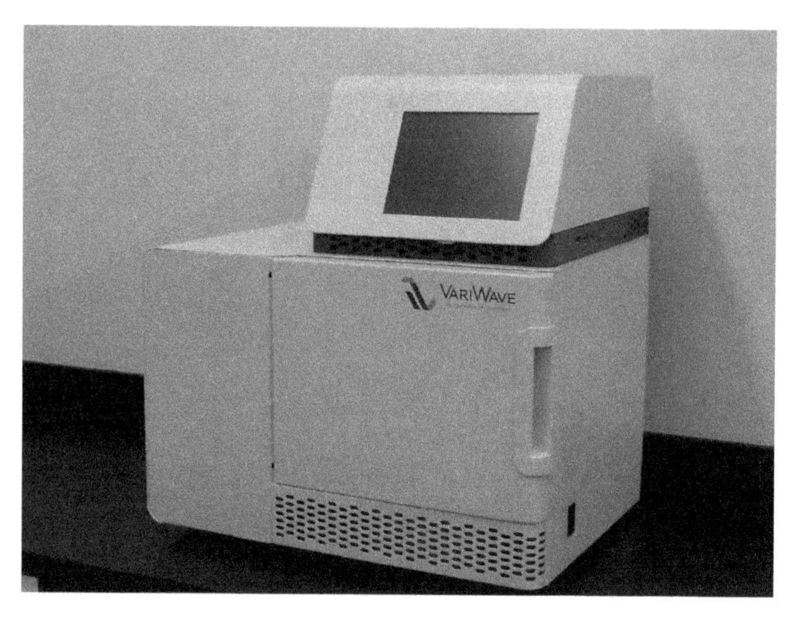

Fig. 5.21 VariWave1200 tabletop VFM system. Reproduced from Lambda Production folder, Copyright 2015 by R. J. Schauer

ratory research and low volume manufacturing of semiconductors, electronics, and medical devices. Figure 5.22 shows the Aurora1330, a broadband tabletop tool integrated with a traveling waveguide applicator that enables unique materials research for specific frequency effects in the presence of catalysts or through plasma exposure. The MASC1330 robotically mounted system shown in Fig. 5.23 was developed to allow delivery of VFM energy to a spot for curing adhesives used to join composites together or to metal parts. Benefits of MASC1330 are fast and selective curing of joining adhesives.

Summary

The use of microwave energy has been found to not only reduce chemical reaction times but also to reduce the temperature of cure necessary for a complete reaction. This technology inherently decreases stress, increases uniformity, improves adhesion, decreases energy consumption, and leads to fewer fracture sites. Further research may well provide enhanced capabilities and perhaps new chemical opportunities.

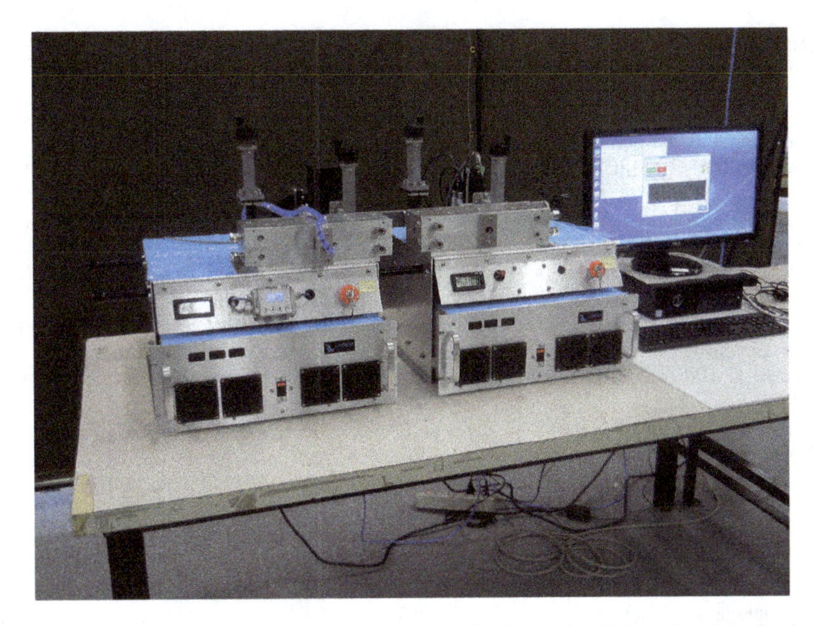

Fig. 5.22 Aurora1330 broadband tabletop single mode. Reproduced from Lambda Production folder, Copyright 2017 by R. J. Schauer

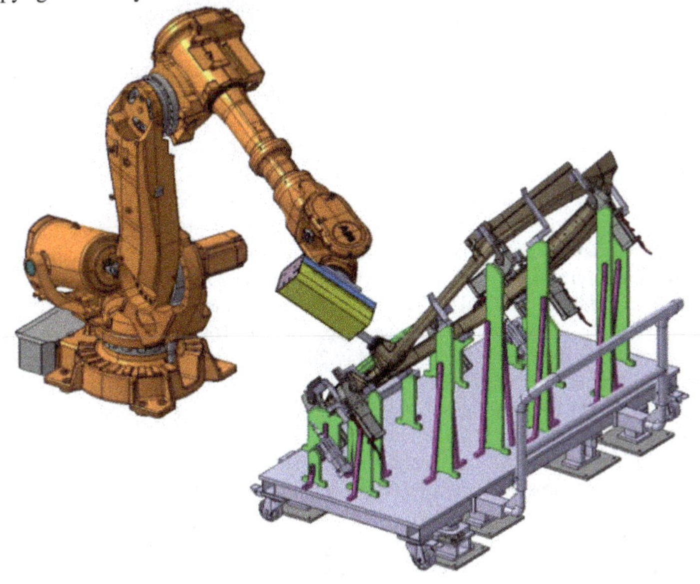

Fig. 5.23 MASC1330 VFM spot curing system. Reproduced from Expert Tooling CAD model library, Copyright 2018 by Andy Bools

References

1. Metaxis AC, Meredith RJ (1983) "Industrial microwave heating", Chapter 2. Peter Peregrinus Ltd., England
2. Lide DR (ed) (1990) CRC handbook of chemistry and physics. CRC Press Inc., Boca Raton, FL, pp 10–193, 10–209
3. Anslyn EV, Dougherty DA (2006) Modern physical organic chemistry, Chapter 1.1. University Science Books, USA, pp 13–19
4. Poly A, Lewis DA, Summers JD, Ward TC, McGrath JE (1992) Accelerated imidization reactions using microwave radiation. J Poly Sci Chem 30:1647–1653
5. Pan B, Nah CK, Chan SL, Wei JB (1998) Variable frequency microwave for chip-on-board glob top curing. In: Proceedings of Nepcon West. Anaheim, CA
6. Tanaka K, Bidstrup Allen SA, Kohl PA (2007) IEEE Trans Compon Packag Technol 30:472–477
7. Hubbard RL, Ahmad I, Hicks K, Zappella P, Brady S, Zemke J, Thiel I, Tyler DR (2010) Reduced warpage and stress in polymers for microelectronics. In: Proceedings of the 14th symposium on polymers in microelectronics. Wilmington, DE, USA, May 2010
8. Hubbard RL, Strain CM, Willemsen C, Tyler DR (2016) Low temperature cure of epoxy thermosets attaining high Tg using a uniform microwave field. J Appl Poly Sci 133:44222–44230
9. Mijovic J, Wijaya J (1990) Comparative calorimetric study of epoxy cure by microwave versus thermal energy. Macromolecules 23:3671–3674
10. Mijovic J, Fishbain A, Wijaya J (1992) Mechanistic modeling of epoxy-amine kinetics. 2. Comparison of kinetics in thermal and microwave fields. Macromolecules 25:986–989
11. Lauf R, Bible DW, Johnson AC, Everleigh C (1993) 2 to 8 GHz broadband microwave heating systems. Microw J 36(11):34
12. Lange J, Altmann N, Kelly CT, Halley PJ (2000) Understanding vitrification during cure of epoxy resins using dynamic scanning calorimetry and rheological techniques. Polymer 41:5949–5955
13. Fathi Z, Tucker D, Ahmad I, Yaeger E, Konarski M, Crane L, Heaton J (1997) Innovative curing of high reliability advanced polymeric encapsulants. In: Proceedings electronic packaging and materials science ix symposium. Boston Materials Research Society, pp 125–135
14. Zussman M, Hubbard RL, Hicks K (2008) Rapid cure of polyimide coatings for packaging applications using variable frequency microwave irradiation. In: Proceedings of the 13th symposium on polymers in microelectronics. Wilmington, DE, USA, May 2008
15. Hubbard RL, Zussman M (2014) Properties and Characteristics of HD4100 PSPI Cured at 250 °C with Microwaves. In: Proceedings of the 16th symposium on polymers in microelectronics. Wilmington, DE, USA, May 2014
16. Hubbard RL, Ahmad I, Hicks K, Ohe M, Kawamura T (2006) Low temperature cure of PBO films on wafers. In: Proceedings of the 12th symposium on polymers in microelectronics. Wilmington, DE, USA, May 2006
17. Hubbard RL, Zussman M (2016) Comparing crosslinking effects in the curing of PBOs. In: Proceedings of the 17th symposium on polymers in microelectronics. Wilmington, DE, USA, May 2016
18. Prime RB (1977) Chapter 6 "Thermosets". In: Turi EA (ed) Thermal characterization of polymeric materials. Academic Press, San Diego
19. Aronhime MT, Gillham JK (1986) Time-temperature-transformation (TTT) cure diagram of thermosetting polymeric systems. Adv Poly Sci 78:83–112
20. Hubbard RL, Zappella P (2011) Low warpage flip-chip under-fill curing. IEEE Trans Compon Packag Manuf Technol 1:1957–1964
21. Lee HL, Neville K (1967) Handbook of epoxy resins. McGraw-Hill, New York, USA
22. Hubbard RL, Tyler DR, Strain CM (2016) The chemistry of low stress epoxy curing. In: Proceedings of the 17th symposium on polymers in microelectronics. International Microelectronics and Packaging Society, Wilmington, DE, USA, May 2016

23. IBM: Diop MD, Paquet MC, Drouin D, Danovitch D (2013) Proceedings 46th international symposium on microelectronic packaging. International Microelectronics and Packaging Society, Wilmington, DE, pp 461–466
24. IBM: Diop MD, Paquet MC, Drouin D, Danovitch D (2015) IEEE Trans Device Mater Reliab 15(2)
25. Taiwan Semiconductor Manufacturing Corporation: US Patent No. 8,846,448
26. Hubbard RL, Ahmad I, Tyler DR (2016) The dynamics of low stress epoxy curing. In: Proceedings of the international surface mount technology association conference. Rosemont, IL, USA, Nov 2016
27. Li L, Xue J, Ahmad M, Brillhart M, Ding M, Lu G, Ho P (2006) Materials effects on reliability of FC-PBGA packages for cu/low-k chips. In: Proceedings IEEE 2006 electronic components and technology conference. San Jose, CA, pp 1590–1594

Chapter 6
RF Cooking Ovens

Christopher Hopper

Abstract The generation of microwave energy by solid-state semiconductors has recently been acknowledged as viable, and perhaps preferable. The reason for this is that solid-state technology allows for the capability for control over electromagnetic properties and characteristics of the microwave system [1]. These properties include the phase, the amplitude, and the frequency. With this amount of control, it is expected, and even demonstrated, that the two main characteristics of microwave generation and delivery system (efficiency and heating uniformity) can be improved significantly. It is not enough, however, to simply control and manipulate the properties mentioned above; there must be a feedback system that would allow this to be done in an intentional manner. There is yet another important difference between magnetron and solid-state technology, namely the ability to measure and respond to forward and reflected power levels. A typical solid-state RF (SSRF) generation system is comprised of a small-signal generation section, a high-power amplifier connected to a heat sink, and a power supply to drive the respective electronics [2]. The feedback, which is typically built into the power amplifier, then allows for the monitoring of the frequency, the phase, and the power levels. With these parameters, and the ability to manipulate them, it is not difficult to imagine that algorithms could be developed that could control these properties in a way which is responsive to the load contained within the cavity.

6.1 Figures of Merit

6.1.1 Resonant Cavities

The rectangular cavity of a microwave oven can be thought of as a waveguide closed by two shorting planes. As such, there are boundary conditions which need to be met on these planes. To satisfy the boundary conditions, the only resonant modes within

C. Hopper (✉)
IBEX, ITW Food Equipment Group, Glenview, IL, USA
e-mail: christopher.hopper@itwfeg.com

S. Horikoshi and N. Serpone (eds.), *RF Power Semiconductor Generator Application in Heating and Energy Utilization*, https://doi.org/10.1007/978-981-15-3548-2_6

the cavity will be those for which an integer number of wavelengths fit between the shorting planes. A good starting point is the wave equation given in Cartesian coordinates (Eq. 6.1) [3, 4],

$$\left\{\frac{\partial^2}{\partial x^2}+\frac{\partial^2}{\partial y^2}+\frac{\partial^2}{\partial z^2}+\omega^2\varepsilon_0\mu_0\right\}\left\{\begin{matrix}E_z\\ H_z\end{matrix}\right\}=0 \tag{6.1}$$

where ω is the resonant frequency, ε_0 and μ_0 are the electric permittivity and magnetic permeability of free space, respectively; E_z is the longitudinal component of the electric field (in V/m) and H_z is the longitudinal component of the magnetic field (in A/m). Supposing that the cavity has dimensions of width b, depth a, and height c, the solution of the wave equation can then be found by the well-known *separation of variables* method given by Eq. (6.2),

$$\left\{\begin{matrix}E_z\\ H_z\end{matrix}\right\}=f(x)g(y)h(z)e^{i\omega t} \tag{6.2}$$

where f is independent of y and z, g is independent of x and z, and h is independent of x and y. Substituting Eq. (6.2) into Eq. (6.1), leads to differential equations with solutions as exponential functions, i.e., $\exp(\pm ik_x x)$, $\exp(\pm ik_y y)$, $\exp(\pm ik_z z)$. Ultimately, the resonant frequencies are found to be given by Eq. (6.3),

$$f=\frac{1}{2\sqrt{\varepsilon_0\mu_0}}\sqrt{\frac{m^2}{a^2}+\frac{n^2}{b^2}+\frac{p^2}{c^2}} \tag{6.3}$$

where m, n, and p are the number of half-wave variations in the x-, y-, and z-directions, respectively. Given values of m, n, and p obviously produce specific field patterns within the cavity, and with the appropriate boundary conditions, the field components can then be solved for analytically. It should be reiterated that this formalism applies to the empty cavity. As shown in Sect. 6.2, the presence of a load will shift mode frequencies and alter the field patterns, depending on the load location within the cavity.

6.1.2 Quality Factor and Mode Separation

The quality, or so-called Q_0 factor of a cavity, is a convenient way of describing how efficient a resonant cavity is. It can be written as the ratio of stored energy U to the power loss P_d as expressed by Eq. (6.4).

$$Q_0=\frac{\omega U}{P_\mathrm{d}} \tag{6.4}$$

When the quality factor is high, Q_0 can also be expressed by Eq. (6.5),

$$Q_0 = \frac{\omega}{\Delta\omega} \tag{6.5}$$

where $\Delta\omega$ is the *bandwidth*. An empty cavity will have three main sources of power loss or dissipation. There is (i) the power lost to the metal walls, (ii) the power reflected back to the source, and (iii) the power leaking from the door.

There is yet another quality factor that is more important to the present discussion: the *loaded* quality factor Q_L, which refers to the quality factor for a resonant cavity that is not empty, or otherwise has additional sources of power dissipation. A loaded cavity has a much lower quality factor, $Q_L \ll Q_0$. This is because, ideally, the load will absorb a large amount of the energy directed into the cavity. This results in a large dissipated power, i.e., Eq. (6.4) becomes small.

6.1.3 Electromagnetic Interaction with Food

There are several ways to transfer energy to a load during cooking or reheating. Those most familiar are conduction, convection, and radiation. Herein, the focus will be on the electromagnetic variety, with radiative heating being just one example. The other two, which will be the subject of this section, are referred to as dielectric and ohmic heating. One difference between the latter two and radiative heating is the frequency of the electromagnetic field; radiative involves infrared, which roughly spans the frequency range 300 GHz to 430 THz. The frequencies used in ohmic heating are generally low but only restricted to the ISM band, while those of dielectric heating are those within this band between a few MHz and several GHz.

Both dielectric and ohmic heating can be considered *direct* heating since the heat is generated within the load as opposed to radiative, conductive, or convective heating where the heat is generated outside of the load and transferred by other means. In ohmic, or Joule heating, the energy from an electric current is directly converted to heat when it encounters resistance in the load. Dielectric heating relies on polar molecules reacting to an alternating electric field. The molecular structure of water, for example, consists of positively charged hydrogen atoms separated from a negatively charged oxygen atom. This structure creates an electric dipole. When polar molecules are in the presence of an applied RF or microwave electric field, they will attempt to orient themselves with the field. The fields are oscillating on the order of 10^6–10^9 times per second (i.e., MHz to GHz), which causes a rapid reversal of molecular orientation that leads to heat generation through friction. It is commonly thought that microwaves heat food from the inside out, but in fact, there is a characteristic depth for which the fields will penetrate. This *penetration depth* is given by Eq. (6.6),

$$\delta = \frac{\lambda_0}{2\pi\sqrt{\varepsilon''}} \tag{6.6}$$

where λ_0 is the wavelength of the RF/microwave radiation and ε'' is the loss factor. The loss factor is one part of the complex permittivity, a quantity that describes the dielectric properties of a material and is given by Eq. (6.7),

$$\varepsilon^* = \varepsilon' - i\varepsilon'' \tag{6.7}$$

where $i = \sqrt{-1}$. The real part of ε^* is the dielectric constant that describes the stored energy when the material is exposed to an electric field. The imaginary part of ε^* is the loss factor, which influences the energy absorption and attenuation within the material. The dielectric properties of materials are affected by many factors such as frequency, temperature, and content of moisture/ionic molecules. The general behavior of the real and imaginary parts in Eq. (6.7) is that the dielectric constant decreases with increasing frequency and temperature, while the loss factor usually has a peak value at a given frequency and decreases with temperature. The real and imaginary parts of the complex permittivity are often combined to characterize dielectric materials with a quantity known as the loss tangent given by Eq. (6.8).

$$\tan\delta = \frac{\varepsilon''}{\varepsilon'} \tag{6.8}$$

Most (biological) materials heated in a microwave have a permeability close to, or the same as, that of free space ($4\pi \times 10^{-7}$ N/A^2) and therefore do not absorb energy from the magnetic field. The power dissipated per unit volume (P_d) in a lossy material from the electric field can be calculated from Eq. (6.9),

$$P_\mathrm{d} = 2\pi f \varepsilon_0 \varepsilon'' |E|^2 \tag{6.9}$$

where $2\pi f \varepsilon_0 \varepsilon'' = \sigma$ is the electric conductivity of the dielectric medium in units of S/m and ε_0 is the permittivity of free space (8.854×10^{-12} F/m). The electric field distribution determines how the applied microwave radiation is converted into power; this absorbed power is the source of heating the material. The equation governing the *diffusion* of heat in a thermally conductive material is given by Eq. (6.10),

$$\rho C_\mathrm{p} \frac{\partial T}{\partial t} = k\nabla^2 T + P_\mathrm{d}(x, y, z, t) \tag{6.10}$$

where ρ is the density (kg/m^3), C_p is the specific heat at constant pressure (kJ/kg °C), and k is the thermal conductivity (W/m °C). It should be noted that in Cartesian coordinates,

$$\nabla^2 = \frac{\partial^2}{\partial x^2} + \frac{\partial^2}{\partial y^2} + \frac{\partial^2}{\partial z^2}$$

In principle, the food would also transfer heat to the surrounding air through convection

$$-\hat{n}(-k\nabla T) = h(T_\infty - T)$$

and radiation

$$-k\hat{n}\nabla T = \left(T^4 - T_\infty^4\right)\sigma_s\epsilon_s$$

where δ_s is the Stefan–Boltzmann constant (5.67×10^{-8} J/s m^2 K^4) and ε_s is the material-specific emissivity. Under normal circumstances, these contributions are negligible. Finally, the volumetric temperature rise can be calculated from Eq. (6.11).

$$\Delta T = \frac{P_d t}{m C_p} \tag{6.11}$$

6.2 Simulations

As mentioned in the previous sections, the electromagnetic field pattern within the cavity is highly dependent on the operating frequency, source-phase relations, cavity dimensions, load geometry, placement, and physical properties. Sophisticated electromagnetic solvers such as COMSOL [5] and CST [6] can effectively help one to understand how this variability can affect the field patterns.

Simulating the empty cavity can provide insight into the modes supported, and respective field patterns produced. However, this information has limited utility when developing solid-state RF technology specifically for cooking since the food can change quite dramatically the electromagnetic field distribution, the resonant frequency, and the mode bandwidth. In addition, the placement of food within the cavity can have a significant effect on these parameters. In both residential and commercial applications (i.e., outside of a laboratory environment), the placement and reproducibility are affected by user choices. This variability is difficult to account for in a simulation.

Given the cavity dimensions, simulations can tell us which modes will be sustained and the relative excitation level. Table 6.1 is an example of simulated and calculated cavity modes along with the Q_0 value of each. The simulated values were found using the RF module in COMSOL version 5.3 [5], and the calculations were carried out using Eq. (6.3). It is clear that many of the available modes are spaced close together in terms of frequency. This leads to overlapping, which is enhanced by the presence of a load since $Q_L \ll Q_0$. Figure 6.1 shows an example of how the field pattern is altered when an ideal (i.e., perfectly symmetric, homogeneous) load is placed in the center of the cavity.

Table 6.1 Calculated and simulated cavity modes

Mode indices			Simulated f_{mnp} [GHz]	Calculated f_{mnp} [GHz]	Q_0
8	2	1	2.404	2.397	33,642
2	6	0	2.405	2.398	37,433
9	0	0	2.407	2.369	31,994
8	3	0	2.415	2.409	33,318
7	4	0	2.421	2.414	34,452
3	4	2	2.430	2.423	41,418
5	3	2	2.439	2.432	42,114
6	2	2	2.439	2.433	43,693
3	6	0	2.476	2.469	37,956
1	6	1	2.492	2.499	27,483
7	0	2	2.499	2.522	34,530
9	2	0	2.506	2.494	33,385

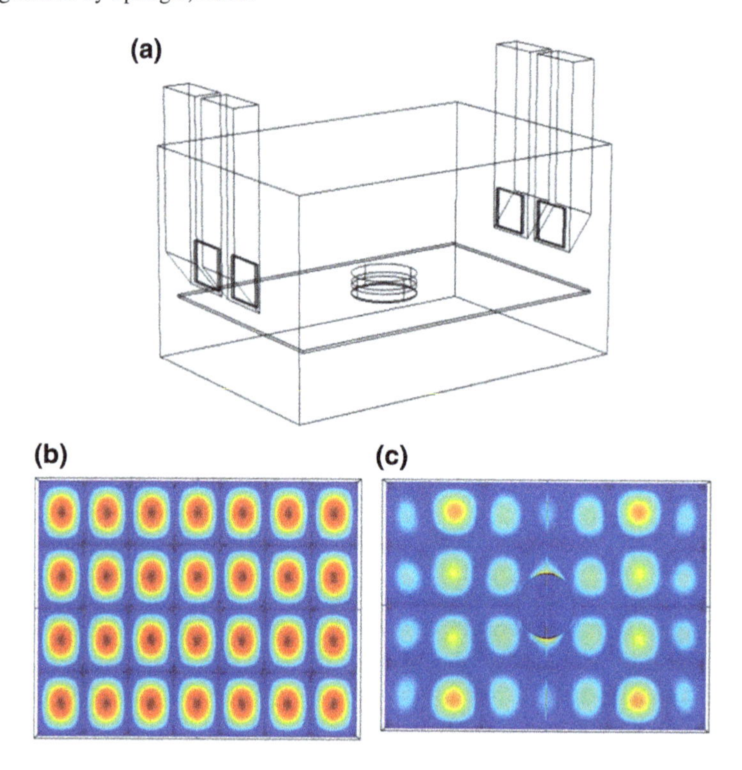

Fig. 6.1 **a** Model cavity with tray and load, **b** empty cavity mode at 2420 MHz, and **c** loaded cavity mode. Copyright 2019 by Springer, Berlin

Another important phenomenon as it relates to microwave cavities is that of standing waves. The modes referred to previously are the resultant interference pattern created in an empty or loaded cavity. An illustration of the phase and resulting wave interference is displayed in Fig. 6.2. The importance of the ability to change phase with two sources is shown in Fig. 6.3. This is a simulation of a full batch where the sources are completely in-phase or completely out-of-phase. This is an extreme case, which can be partially understood from the simple 2D depiction illustrated in Fig. 6.2. When the waves are in-phase (Fig. 6.3a), a very regular pattern of hot and cold spots is produced. In the other extreme (Fig. 6.3b), there is much less power

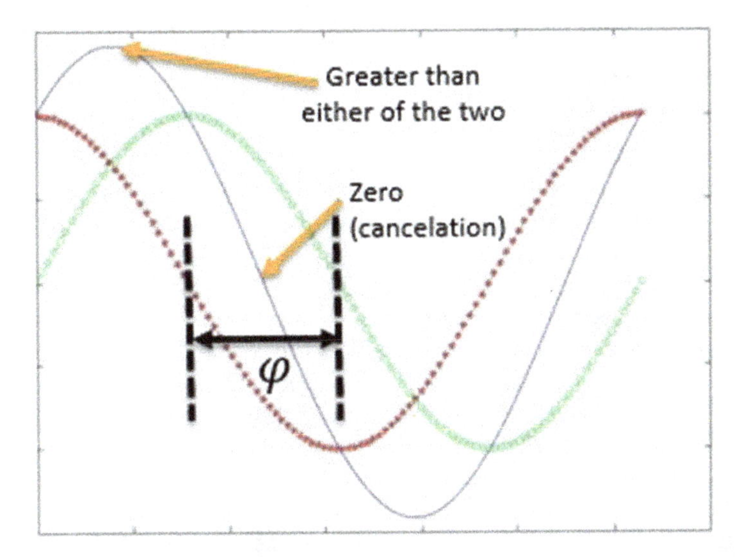

Fig. 6.2 Depiction of the phase difference between two waves and the resulting linear combination. Copyright 2019 by Springer, Berlin

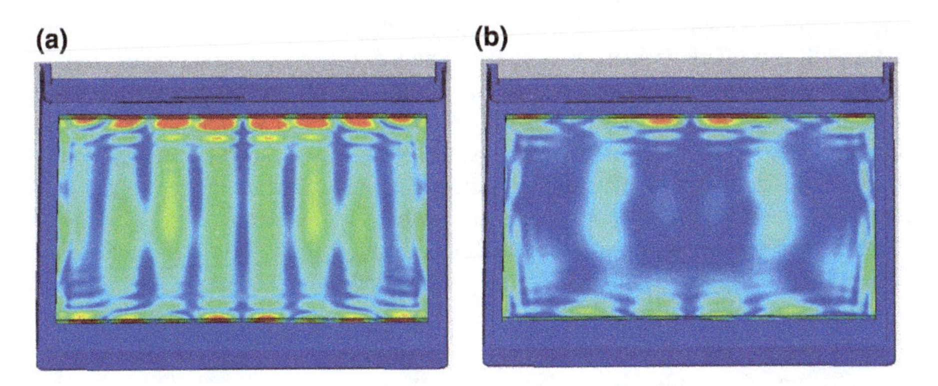

Fig. 6.3 Two 2450 MHz (i.e., 2.45 GHz) sources operating with a phase difference of **a** 0° and **b** 180°. Copyright 2019 by Springer, Berlin

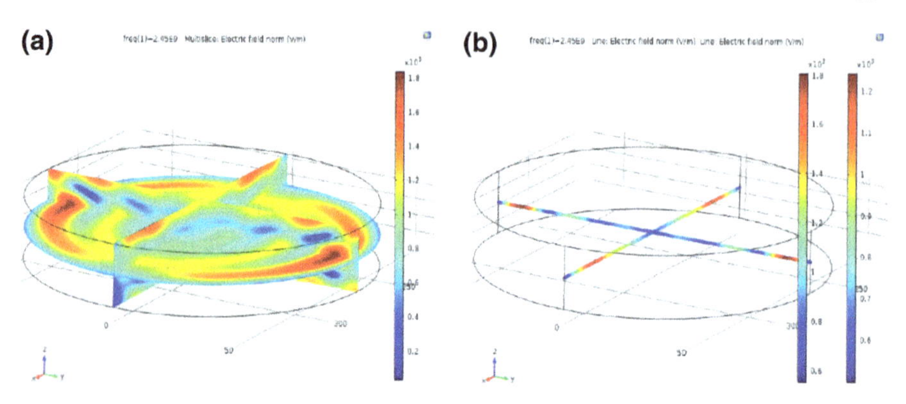

Fig. 6.4 **a** Normalized electric field on *x*-, *y*-, and *z*-slices and **b** the normalized electric field through the center of the load. Copyright 2019 by Springer, Berlin

being dissipated in the load owing to destructive interference. Having all the combinations between these extremes can produce much more uniformity in heating than either one individually.

When loaded, this interference pattern in the food accounts for the familiar hot and cold spots produced when heating or cooking in a microwave cavity. As before, simulations can be useful in understanding the expected field pattern under different operating conditions. The interference of the various Cartesian components of the electric field can also be estimated to better understand the resulting heating pattern that results from resonances within the load itself [7]. For example, Fig. 6.4b shows lines passing through the center of a cylindrical load, which can be used to visualize how the field components change with varying parameters. Figure 6.5 is an example of the Cartesian field components evaluated along these lines.

With this type of simple visualization, we can better understand which field components are responsible for heating in a given area of the load. In other words, evaluation of the interference pattern within a load can help us understand the hot and cold spots which are typical when heating with a microwave oven. Figures 6.5b, e, f are all zero. This tells us $E_y(x)$, $E_y(y)$ and $E_z(y)$ are either naturally zero or interfere destructively within the load. In addition, $E_x(x)$ and $E_x(y)$ are large and interfere to produce an even mode within the load. $E_z(y)$ has a smaller magnitude and produces an odd mode interference pattern. The way these field patterns can be interpreted is consistent with what was stated previously. The component of the field parallel to the surface is responsible for the heating. In this simple case, we see that there is a strong *x*-component along both the *x*- and *y*-axes. This is parallel to the edges with the highest heating, as would be expected.

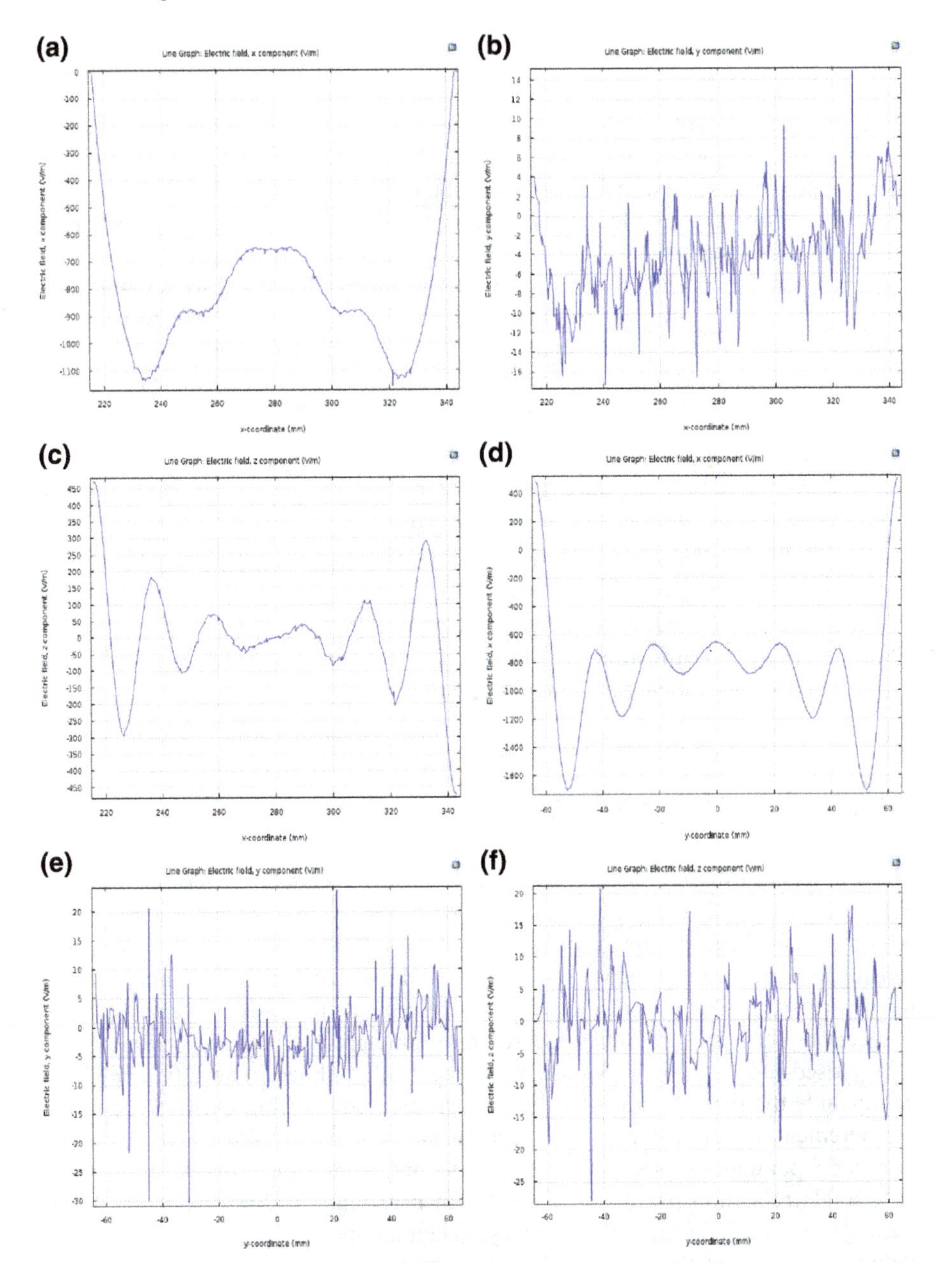

Fig. 6.5 **a** $E_x(x)$, **b** $E_y(x)$, **c** $E_z(x)$, **d** $E_x(y)$, **e** $E_y(y)$, and **f** $E_z(y)$. Copyright 2019 by Springer, Berlin

6.3 Intelligent Cooking

In the context of a solid-state RF oven, intelligence can be roughly defined as the ability to obtain relevant feedback from the system and respond in a way which appropriately respects the desired outcome of the user, e.g., uniform, efficient heating. Under this definition, it is clear that techniques such as mode stirrers and mechanical turntables used in conventional microwave ovens are not "intelligent."

In order to be considered intelligent, the device needs to acquire feedback and has the ability to change variables in response. Having access to the forward and reflected power at each power amplifier (PA) is sufficient for simple applications. This is the feedback which can change during the cooking process. Some parameters that can be used to respond to the feedback are:

- Frequency bandwidth spanning several tens of MHz.
- The phase shift between each RF source with respect to a reference. For this variable, it is the difference in phase between two sources that matters.
- The amount of time spent with the power on at each frequency/phase combination.
- The amplitude of the output power per source.

Having access to the forward and reflected power coupled to the ability to manipulate the variables outlined above allows for a truly intelligent appliance that can monitor and adjust the energy being deposited in the load.

6.4 A Closed-Loop System

Many magnetron-based cooking appliances can be considered open-loop systems. This means that, if left unattended, they will attempt to output the same power with the same field distribution within the cavity. There are some, however, with sensing technology not directly related to the cooking electronics. For example, some microwave and RF devices can sense the moisture content in the cavity and adjust the magnetron "on" and "off" times according to some predetermined functions [7, 8].

A closed-loop system is enabled by the feedback outlined in Sect. 6.3. Feedback can include the forward, reflected, and transmitted power from each source. These measurements are often made with a vector network analyzer (VNA) and produce the familiar S-parameters used to characterize the matching of a network. An example of a single reflection parameter (S_{11}) is shown in Fig. 6.6. Specifically, a closed-loop system works in the following way. When an operator requests RF energy be delivered to the load, the oven scans a predetermined number of phase configurations for a discrete number of frequencies. These data are collected and parsed depending on the algorithm being employed. When certain criteria are satisfied, the oven will deliver RF energy to the cavity for an amount of time. This process continues until the load is cooked.

An example can be useful to illustrate how important a closed-loop system can be. Suppose we are heating a load with a high water content which has been marinated

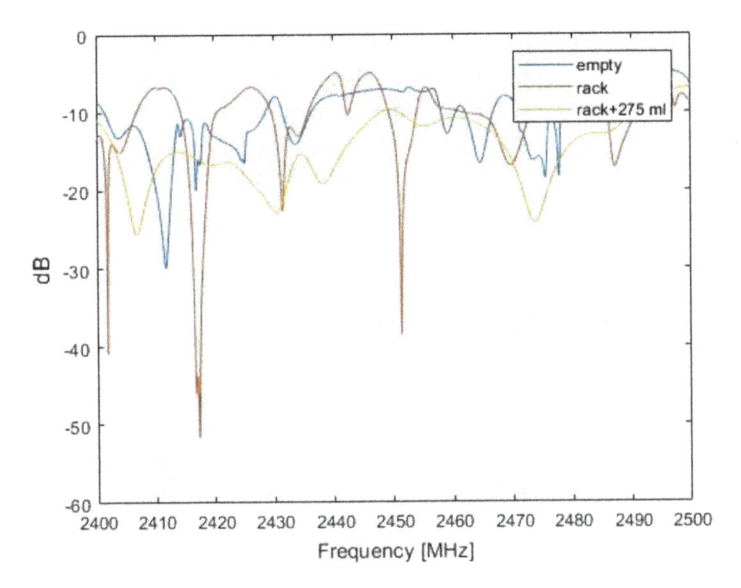

Fig. 6.6 S_{11} with and empty cavity, cavity with generic cooking platform, and with the cooking platform and a water load. Copyright 2019 by Springer, Berlin

in a salty mixture. Water molecules are partly responsible for the interaction with the RF fields, and salt adds ionic conduction. For water, the way the parameters ε' and ε'' in Eq. (6.7) change as a function of temperature (and salinity) is shown in Fig. 6.7 [9–11].

The dielectric constant can change by 20% or more when going from ambient to 80 °C, while for very salty products, the loss factor can more than double. One can imagine that the reflected power read by the electronics could change significantly during this process. If the configuration of the system is not reconsidered throughout the cooking process in order to adapt to the changing physical properties of the load,

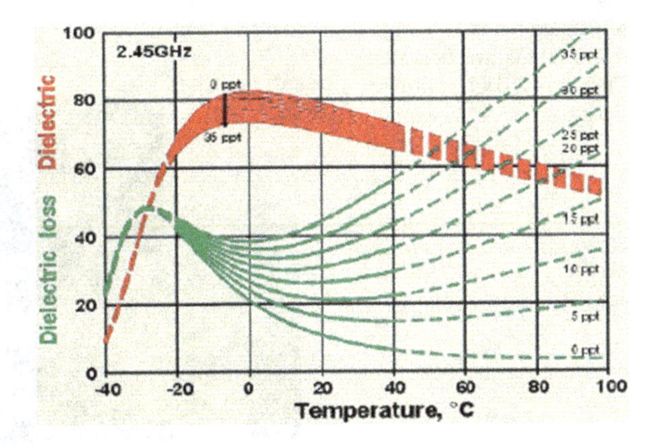

Fig. 6.7 Dielectric properties of water. Reproduced from Ref. [9]. Copyright 2019 by Springer, Berlin

hot spots can turn into heat sinks, which can exacerbate any underlying nonuniformity present. Being able to adapt to these and other changes in the load as it is heated can have a significant impact on the outcome. In addition, recent work [12] has shown promising results which utilize this phenomenon to address user requirements, such as location determination of the load, cookware identification, and even temperature prediction.

6.5 Performance Characterization

The ability to achieve an improved performance in a solid-state cooking oven has been extensively written about. To date, there is limited experimental demonstration of this. The following section will present results obtained using the IBEX ONE commercial solid-state microwave/convection oven pictured in Fig. 6.8. Other solid-state ovens have been proposed and/or are available in the residential space [13–16].

6.5.1 Efficiency

Efficiency itself does not tell us much about the quality of the food produced in a particular type of oven. Standardized testing of efficiency and uniformity has been developed and extensively used for magnetron-based ovens [17, 18]. These techniques can be useful for comparing efficiency between magnetron and solid-state ovens, in particular on how the size and shape of the load change the efficiency. Calorimetric tests are useful for efficiency evaluation of water. For example, the ability to change phase and frequency in order to improve impedance matching with the load makes is possible to maintain a high level of efficiency for various water

Fig. 6.8 IBEX commercial solid-state microwave/convection oven. Copyright 2019 by Springer, Berlin

Table 6.2 Efficiency of solid-state and magnetron-based rapid cook ovens

IBEX			
Efficiency	Container type	T1 [°C]	T2 [°C]
High	Square box	22.4	50.3
High	Sheet pan (plastic)	22.7	48.5
High-med	Small cup	22.6	70.1
Magnetron-based			
Efficiency	Container type	T1 [°C]	T2 [°C]
High	Square box	21.4	71.3
High	Sheet pan (plastic)	22.1	64
High-med	Small cup	22.4	86.6

configurations. Table 6.2 lists some generalized efficiency results obtained using Eq. (6.12),

$$\eta = \frac{mC_{\mathrm{p}}\Delta T/t}{P_{\mathrm{fwd}}} \tag{6.12}$$

where m is the mass of the load(s), C_{p} is the specific heat capacity of water, ΔT is the change in temperature, t is the heating time in seconds, and P_{fwd} is the total *requested* forward power. The requested power can be multiple kW in a magnetron-based oven, whereas solid-state ovens currently have less than that. This explains how the final temperature can be high while having a low efficiency rating. More sophisticated methods for assessing heating performance have also been done [19] but will not be detailed here.

The results in Table 6.2 can give an idea of how quickly a food item can be heated in the ovens. However, caution should be taken when trying to extrapolate from simple RF heating of water to actual cooking. Using a multicomponent sandwich along with the rapid cook oven (magnetron-based) manufacturer's suggested recipe, the result was compared to the solid-state oven (IBEX) which can maintain a high efficiency even with small loads. Table 6.3 shows the results for two different sandwiches.

6.5.2 Uniformity

Heating uniformity certainly contributes to the overall food quality, and simulations can often be useful for quantifying the effects of using solid-state sources [1, 20] versus their magnetron-based counterparts. Thermal imaging and internal temperature monitoring can help validate the simulations and quantify the heating uniformity. This validation process is extremely important when developing cooking algorithms which take advantage of the closed-loop systems available with solid-state ovens

Table 6.3 Comparison of heating sandwiches in IBEX and magnetron-based rapid cook ovens

Quantity	IBEX	Magnetron-based
Sandwich 1		
Mass	270 g	270 g
Ti	8 °C	8 °C
Tf, hot	61 °C	92 °C
Tf, cold	42 °C	43 °C
Time [mm:ss]	00:45	01:25
Sandwich 2		
Mass	1050 g	105 g
Ti	8 °C	8 °C
Tf, hot	58 °C	72 °C
Tf, cold	45 °C	20 °C
Time [mm:ss]	01:45	02:40

[21]. Figure 6.9 shows some plots made from experimental data and the associated simulation results for a single frequency with two different phase configurations (called ϕ_1 and ϕ_2). The load consists of water in a silicon tray heated in the IBEX oven. It is clear that while the simulation is not a perfect match with the experimental

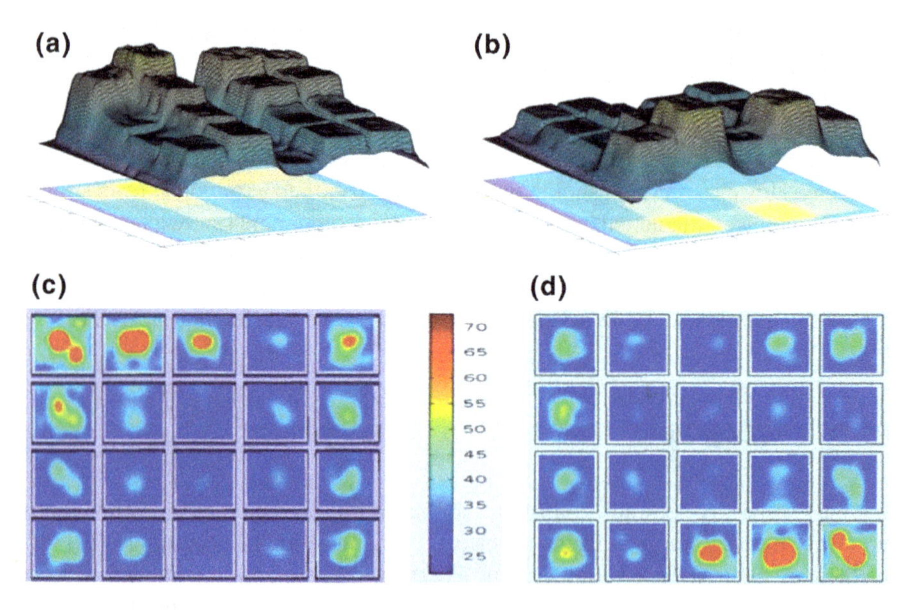

Fig. 6.9 Heating profiles for two different phase configurations **a** experimental result with ϕ_1, and **b** ϕ_2. **c** Simulation results for ϕ_1 and **d** ϕ_2 Copyright 2019 by Springer, Berlin

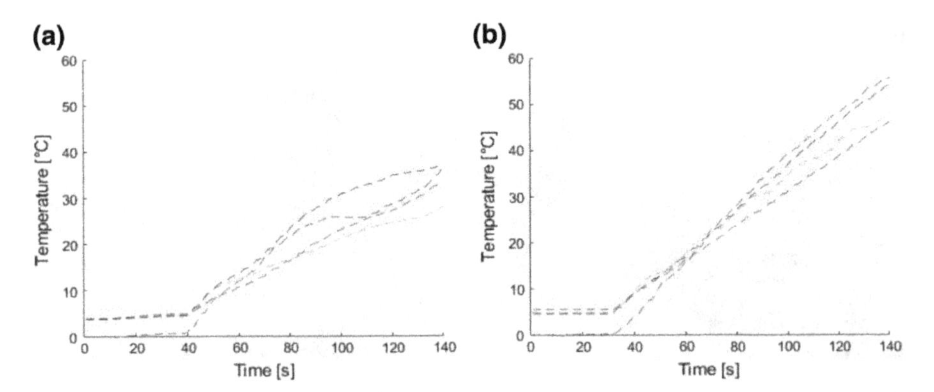

Fig. 6.10 Temperature increase over time for the inner component of a reheated sandwich. In **a**, the frequencies and phases are not optimized to target this component, and in **b**, they are. Reproduced from Ref. [12]. Copyright 2019 by Springer, Berlin

results, they do show a recognizable heating pattern. Clearly, using a combination of ϕ_1 and ϕ_2 would produce a more uniform final result than either one individually.

In addition to being able to improve uniformity by intelligently combining different field patterns, the control available with solid-state technology also allows for the targeting of specific components within a multicomponent product or certain discrete items within the cavity. For example, in certain commercial sectors, reheating products like pastries and sandwiches are common. Sandwiches, in particular, often have a denser inner component surrounded by some type of bread. In an oven which combines convection and microwave heating, the microwaves are generally not needed to heat the bread in the required time. In this case, the microwaves would ideally optimize the heating rate of the inner components. This is possible with a closed-loop system. Figure 6.10 shows an example of a general heating strategy which has not been optimized to heat the internal egg and the result after applying an optimized algorithm. Both measurements were taken while applying low power to extend the overall heating time. When comparing Fig. 6.10a, b, it is clear that having control over the phase and frequency alone does not necessarily produce improved results over what could be expected from a magnetron-based system. In fact, the increased efficiency and stability of the solid-state RF (SSRF) system could produce worse heating uniformity if not used in the proper way [22]. For cooking applications, additional challenges arise as a result of varied component properties, inconsistency in manufactured products, nonsymmetric geometries, and item placement. In other words, repeatability and reproducibility are not automatically achieved [23] without sophisticated cooking algorithms. Additionally, statistical significance testing is needed in order to quantify confidence levels in the results. Maximizing power transfer to the cavity [24] is often the goal, but in cooking applications, it is sometimes necessary to maximize the power transfer to *specific* components of the load, which may be a local, rather than global, maximum.

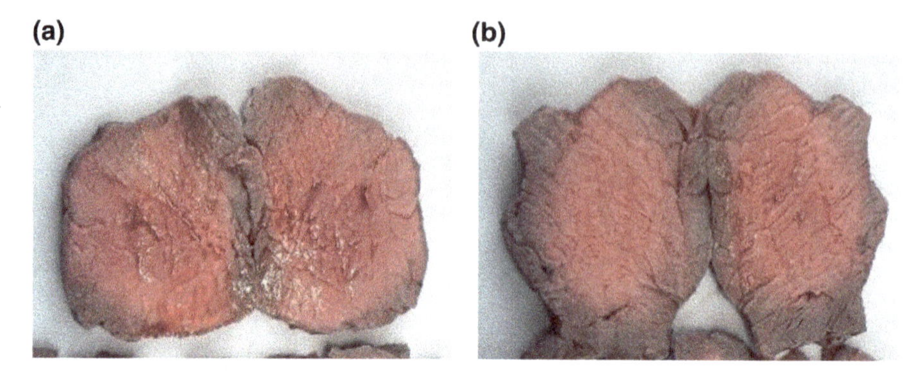

Fig. 6.11 Cross section of a beef sirloin cooked in a rapid **a** solid-state oven and **b** a magnetron-based cook oven. The product mass and cooking time were the same for both ovens prior to cooking. Copyright 2019 by Springer, Berlin

Uniformity also applies to single-component loads such as large-mass proteins. In this case, minimizing or eliminating hot spots is extremely important. Low microwave power may be more desirable, and with the proper recipe and algorithm(s), the final results are both quantitatively and qualitatively improved. Figure 6.11 shows an example of a beef sirloin cooked for the same amount of time in the IBEX and a magnetron-based rapid cook oven.

Proteins and raw doughs are challenging items to build a general-purpose algorithm for, as there are chemical and structural changes taking place during cooking that require the system to adapt in an informed way. When baking, a lack of uniformity in the electromagnetic field distribution can also lead to nonuniform browning through convection.

6.6 Concluding Remarks

Solid-state RF ovens are becoming a reality, both in the commercial and residential sectors. A successful adoption of this technology will depend on several factors. First, being able to produce the entire RF electronics and control systems in a cost-effective manner is the key for reducing the product price and will enable more equipment manufacturers to develop their own solutions. Secondly, the lifetime, reliability, and repeatability have to be demonstrated and documented in the intended operating environment. Perhaps most importantly, investment in resources needed to advance our understanding of what is possible using the closed-loop systems solid-state RF allow.

References

1. Yakovlev VV (2016) Computer modeling in the development of mechanisms of control over the microwave heating in solid-state energy systems. Ampere Newsl 89:18–21
2. Werner K (2015) RF energy systems: realizing new applications. Microw J: 22–34
3. Waldron RA (1970) Theory of guided electromagnetic waves. Van Nostrand Reinhold Company, London
4. Jackson JD (1999) Classical electrodynamics, 3rd edn. John Wiley & Sons Inc., Hoboken
5. COMSOL Multiphysics, v. 5.3. www.comsol.com
6. Computer simulation technology. www.cst.com
7. Datta A (2001) Handbook of microwave technology for food application. Marcel Dekker Inc., New York
8. http://products.geappliances.com/appliance/gea-support-search-contentID=21531
9. www1.lsbu.ac.uk/water/microwave_water.html
10. Ellison WJ (2007) Permittivity of pure water, at standard atmoshpheric pressure, of the frequency range 0–25 THz and the temperature range 1–100 C. J Phys Chem Ref Data 36:1
11. Ellison WJ, Lamkaouchi K, Moreau JM (1996) Water: a dielectric reference. J Mol Liq 68:171279
12. Hopper CS, Grimaldi G, Marra F, Prochowski J, Sclocchi M (2019) Targeted food heating using a solid-state microwave oven. In: Proceedings of the 53rd annual microwave power symposium. Las Vegas, Nevada, USA
13. https://revolutionaryexcellence.miele.com/en/dialog-oven
14. https://www.ampleon.com/news/press-releases/ampleon-and-midea-collaboration-results-in-world-s-first-solid-state-oven.html
15. https://www.nxp.com/products/rf/rf-power/rf-cooking/smart-cooking-concept:RF-SMART-COOKING-PG
16. http://www.gojifoodsolutions.com/rf-cooking-technology
17. Laug OB (1977) Evaluation of a test method for measuring microwave oven cooking efficiency, NBSIR 77-1387. Institute for Applied Technology, National Bureau of Standards, Washington D.C.
18. Household microwave ovens-Methods for measuring performance. International Electrotechnical Commission, IEC 60705, Geneva, Switzerland (2010)
19. Pitchai K, Birla S, Suggiah J, Jones D (2010) Heating performance assessment of domestic microwave ovens. In: 44th annual symposium of the International Microwave Power Institute
20. Yakovlev VV Effect of frequency alteration regimes on the heating patterns in solid-state-fed microwave cavity. J Microw Power Electromagn Energy 52(1):31–44. https://doi.org/10.1080/08327823.2017.1417105
21. Hopper CS, Grimaldi G, Mannara G, Marra F, Sclocchi M (2018) Analysis of food heating in a multi-source solid-state microwave oven. In: 52nd annual symposium of the International Microwave Power Institute
22. Wesson R (2016) Solid-state microwave cooking. AMPERE Newsl 89
23. Werner K (2016) RF energy alliance: advancing next generation microwave heating applications. AMPERE Newsl 89
24. Durnan G (2016) Solid-state heating with advanced RF-power solutions. AMPERE Newsl 89

Chapter 7
Radio Frequency (RF) Discharge Lamps

Stephan Holtrup, Satoshi Horikoshi and Nick Serpone

Abstract This chapter guides the reader through a brief historical view of discharge lamps, especially RF discharge lamps, from their first accounts in the late nineteenth century until novel developments. Some principles of novel lamp technologies are explained, and their advantages pointed out in several special applications apart from the well-known general lighting.

7.1 The Beginning

If you ask people about the inventors of modern artificial lighting, Thomas Edison is surely the one name you will often come across. Many inventors worked on the incandescent light bulb around 1880. For instance, Sir Joseph Swan was also a pioneer in this field. He used carbonized paper as a filament in a partially evacuated bulb that lasted several hours. This was a great achievement compared to the scores of inventors who could not get incandescence to last [1]. However, with his base 1889 patent (No. 223.898 [2]), Edison developed a method to manufacture carbon filament lamps successfully, which lasted for about 800 h [1]. After his invention of the incandescent light bulb, 27 years later (1907) Edison further invented and patented one of the fluorescent lamps. His coating on the glass tube fluoresced when subjected to X-rays that were generated inside the vacuum tube [3]. Subsequent to the passing of Clarence Dally as a result of having been exposed to X-ray radiation, Edison abandoned the project with the consequence that his version of the fluorescence lamp never went into production [4].

The first observations of a glow discharge were made nearly 50 years before Edison's 1889 patent. Indeed, in 1856, the German artist and glassblower Heinrich

S. Holtrup (✉)
pinkRF, Transistorweg 7d, 6534 AT Nijmegen, The Netherlands
e-mail: stephan.holtrup@pinkrf.com

S. Horikoshi
Department of Materials and Life Sciences, Sophia University, Tokyo, Japan

N. Serpone
PhotoGreen Laboratory, Dipartimento di Chimica, Universita di Pavia, 27100 Pavia, Italy

S. Horikoshi and N. Serpone (eds.), *RF Power Semiconductor Generator Application in Heating and Energy Utilization*, https://doi.org/10.1007/978-981-15-3548-2_7

Geissler was the first to examine arc tubes. He observed the glow of air in a tube at low pressures while applying an alternating current under high-voltage conditions. Accordingly, the Geissler Tube was the first groundwork for all subsequent arc discharge lamps [4]. He and other inventors, such as Faraday and Crookes (among others), indicated that all individual gases or vapors would carry a current, with some giving off a fairly strong light [5].

Work on discharge lamps continued with Daniel McFarlan Moore at the end of the nineteenth century. By 1898, Moore had devised his "Moore Lamp" (Fig. 7.1) [6] that consisted of gas-filled glass tubes approximately 2 inch in diameter joined together in lengths up to 250 ft. Once installed, air was removed from the tubes and small amounts of gases, typically nitrogen or carbon dioxide, were introduced, after which an electric current was passed between the electrodes mounted at either ends of the lamp (in a manner similar to later neon tubes inspired by Moore's work). Electric current and gas pressure were regulated by devices installed in a box from which both ends of the tube emerged [7]. Nonetheless, a problem with Moore's lamp was that it was not modular and flexible to interior spaces and therefore could not be adapted to changing needs.

After Moore's development of the first commercial arc tube in 1895, Peter Cooper Hewitt developed one of the first mercury vapor lamps in 1901 [8]. In his 1901 patent he claimed a method to produce mercury-filled gas discharge lamps (Fig. 7.2) produced in standardized sizes and needed voltages lower than required in Moore's lamps. The lamps emitted a blue-greenish light and were portable. Unfortunately, they also contained a non-insignificant quantity of mercury (1 lb = 454 g) [8]. For the lamp to ignite, the tube had to be tilted so that the quantity of mercury shorted

Fig. 7.1 Moore's lamp installed in a gallery in 1906. For greater details regarding this figure, see: http://edisontechcenter.org/DFMoore.html (accessed October 2019)

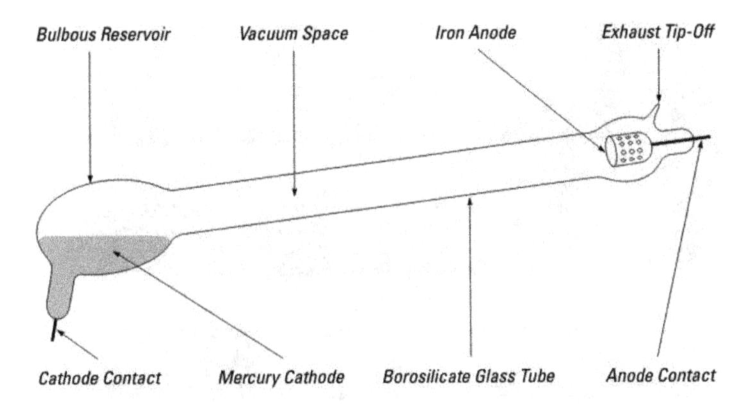

Fig. 7.2 Illustration of the Cooper Hewitt's mercury lamp in the early twentieth century; the lamp tube is tilted so as to cause ignition. For greater details regarding this figure, see: http://www.lamptech.co.uk/Documents/M6%20Cooper%20Hewitt.htm (accessed October 2019)

the two electrodes and caused some of the mercury to evaporate when the electric current passed through the lamp. Upon returning the lamp to its normal position, the mercury moved back into its reservoir, thereby freeing the electrode and producing an arc. This mechanism precluded the usage of high voltages as needed in Moore's lamps, since the latter were filled with nitrogen and carbon dioxide.

Later on, in 1902, the French engineer and chemist George Claude developed a process to harvest inert gases such as neon and proceeded to use it in Moore's lamp. He applied an electrical discharge to a sealed tube with neon as the gas, thereby creating the first neon lamp. However, it was not until 1912 that the first commercial neon lamp was installed for the first time in a barber shop in Paris [10]. This neon-type tube was the original light source that was finally employed into the first commercial fluorescent lamps launched by General Electric in 1938 (Fig. 7.3) [9]. The 14 W version was filled with argon and mercury. It created a luminous flux of 490 lm and lasted about 1500 h.

Despite Moore's lamp principle that formed the basis of fluorescent lamps known from our modern daily life, it was Hewitt who patented the first operational principle of an induction lamp in 1907 in generating an electric light [11]. Three years later, he was granted a patent for his induction lamp [12]. This invention was recorded much earlier than the first commercial electrodeless lamps that ever came commercially available. The timescale of the incandescent light bulb and its history of more than a hundred years are now commonly known. By comparison, the arc-, glow-, and radio frequency (RF) discharge lamps have a similar long history, at least insofar as the ideas are concerned.

Fig. 7.3 First commercial 14 W fluorescent lamp launched by General Electric in 1938. For greater details regarding this figure, see: http://www.lamptech.co.uk/Spec%20Sheets/D%20ED%20Osram%20Endura.htm (accessed October 2019)

7.2 Radio Frequency (RF) Discharge Lamps

The commercializing of radio frequency (RF) discharge lamps began in the early 1990s. A first demonstrator of the Philips quartz lamp (QL) was presented in 1985 and an improved version in 1992 (Fig. 7.4) [9]. This type of lamp is based on inductively

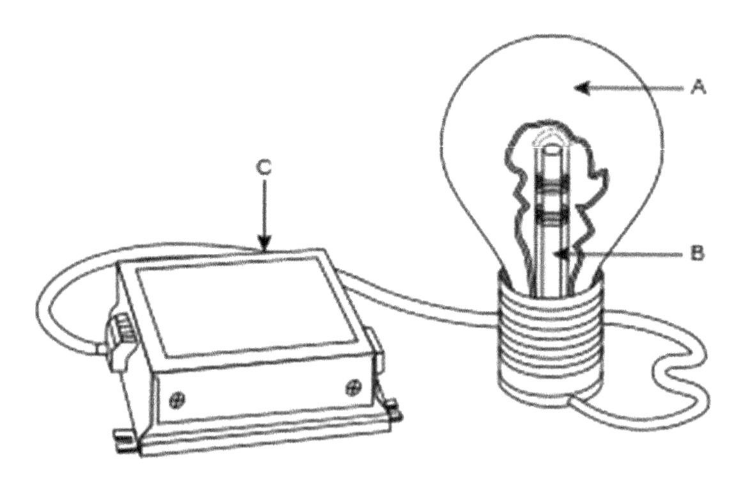

Fig. 7.4 Philips QL lamp system showing the central coil (B) surrounded by the low-pressure vessel (A) including the ballast (C). Figure and related text reproduced from https://en.wikipedia.org/wiki/Electrodeless_lamp (Ref. [14]) under the Creative Commons licenses on Wikipedia and Wikimedia Commons (accessed October 2019)

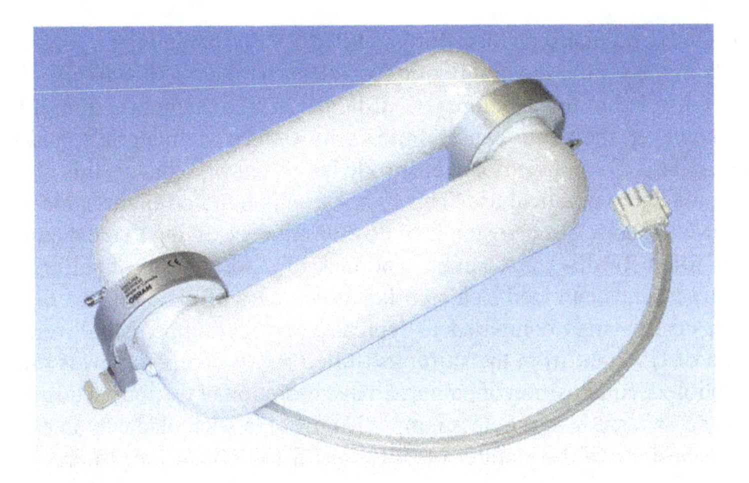

Fig. 7.5 Osram Sylvania's Endura lamp with the two transformers around the closed-loop electrodeless bulb. For greater details regarding this figure, see: http://www.lamptech.co.uk/Spec%20Sheets/D%20ED%20Osram%20Endura.htm (accessed October 2019)

coupled plasma at a frequency of 2.65 MHz with no electrodes present in the low-pressure gas volume. The vessel was filled with 70 Pa of a rare gas (e.g., krypton) and a quantity of mercury (ca. 6 mg) [13]. As illustrated in Fig. 7.4 [14], the vessel surrounds a rod-shaped coil to which a high-frequency signal is applied to. The coil generates a magnetic field in the vessel and forms the inductively coupled plasma. The outer part of the vessel is coated with a conductive and phosphorus coating that prevents the lamp from emitting RF radiation and converts the 254-nm UV radiation into visible light. This conversion process is also used in conventional fluorescent lamps. The commercial version of the QL light source became available around 2001.

Osram Sylvania launched a slightly different electrodeless lamp soon after the QL lamp was launched. The Osram Endura lamp (Fig. 7.5) [9] is filled with krypton as a buffer gas and mercury as the main illuminating element. Since mercury was used under low-pressure conditions, the monochrome 254-nm UV light was converted into visible light by the phosphorous coating on the inner wall of the vessel. The electrical energy is transferred into the plasma via the transformer rings around the vessel using a frequency in the higher kHz region [15].

A different lighting approach was the sulfur-based lamp invented by Michael Ury and his colleagues as they worked for fusion lighting [16]. The sulfur lamp was an electrodeless high-power light source using sulfur as the illuminating element under high pressure conditions. The outstanding characteristic of this lamp was its solar-like spectrum. Molecular sulfur S_2 creates a broad spectrum under high pressure conditions. Johnston developed a self-consistent local thermal equilibrium (LTE) model for the sulfur lamp. The lamp is filled with 1 bar Ar and for a pressure of 5 bar in S_2 he found S_2^+ as the dominant ion at plasma temperatures above 5000 K [17]. RF discharge lamps filled with sulfur gas for illumination purposes have been reported

[17–20]. This particular type of sulfur gas-filled RF discharge lamps generates light principally in the visible spectral region (wavelength range: 370–840 nm). Research in this area has led to the development of lamps that can simulate sunlight. Unfortunately, however, the sulfur can corrode various electrode materials, and thus has proven unsuitable for conventional electrode lamps. At first glance, this observation may not have led to practical use. This distraction is not a problem in ***electrodeless lamps***, however, because corrosion is of no consequence owing to the nature of such sulfur gas-filled RF discharge lamps. The lack of deterioration of sulfur gas-filled RF discharge lamps can lead to a significant decrease of maintenance time and to significant cost savings compared to typical electrode lamps. Moreover, because generation of UV light from the sulfur gas-filled RF discharge lamps is minimal, it prevents photodecomposition of paints and discoloration of exhibits. Thus, the use of microwave discharge electrodeless lamps (MDELs) in such places is an example of making the best use of the features of sulfur gas-filled RF discharge lamps. Of some interest, such lamps have been considered, but not used, for potential application as a light source whenever simulated sunlight is needed for growing plants as on the international space station, for example. Additional applications of such lamps are found in aircraft hangars, in interior illumination of buildings, and in artificial solar lamps in industrial plants. The degree of ionization of the $\boldsymbol{S}_2$ sulfur was of the order of 10^{-5}. In this regard, it is worth noting that sulfur gas-filled RF discharge lamps are currently being used in lighting the Smithsonian museums in Washington D.C. (USA), an example of which is illustrated in Fig. 7.6 (see also Ref. [21]).

Fig. 7.6 **a** Example of a sulfur-filled microwave discharge electrodeless (MDEL) light source; **b** lighting of the Smithsonian Aerospace Museum by a sulfur-containing microwave discharge electrodeless lamp. Photos provided through the courtesy of Fusion UV Systems, Japan KK (see also Ref. [21])

7.3 Attractive Features of RF Discharge Lamps

RF discharge electrodeless lamps powered by microwaves include a surface-wave discharge system in which the electrodeless lamp is installed in a waveguide in addition to a general one in which an electrodeless lamp is installed in a microwave applicator. In the surface-wave discharge, microwave power is transmitted through the boundary surface of the discharge plasma in the lamp. For this reason, the light grows longer when power is increased and becomes shorter when power is decreased. Suffice to note that the electrodeless discharge lamp has the following characteristics.

1. Constant light output without being affected by ambient temperature.
2. The circuit is small for high-frequency lighting, and the light output is easy to control.
3. It has a long service life requiring only a few lamp maintenances cycles and is suitable for places where maintenance is difficult.
4. Light color can be set arbitrarily, and light control and color control are easy to achieve.
5. Since the arc tube is small, the light has high brightness and is highly efficient; light control is easy and the equipment can be made small.
6. Since there are no electrodes, the range of gas inclusions can be selected, and a new light source can be obtained in terms of light color efficiency.
7. Since no electrodes are used, a new luminescent material can be used that could not be used with electrode materials.
8. There are no energy losses at the electrodes as the lamps are electrodeless, and thus the lamp is highly efficient.

7.4 Novel Application of RF Discharge Lamp

Different types of microwave discharge electrodeless lamps (MDEL) are currently being used in various industrial processes depending on the type of applications as, for instance, in paint hardening (dryness) equipment [22], a technology that was developed in the USA in the early 1970s to cast a polymer rapidly by means of a UV hardening method starting from a monomeric material. Usage of microwaves is best in such applications. Note that a MDEL light source is a kind of microwave-driven RF discharge lamp. The process yields a thin film of variable thickness compared to the more conventional drying process. Moreover, volatile organic compounds (VOCs) generated from the evaporation of organic solvents can be controlled by this system. The MDEL emits a stable light irradiance for long time periods so that the UV hardening method with an MDEL system fits nicely in an industrial process. At present, MDELs are widely used, among others, in the printing process, in paints, in coatings of a DVD surface, and in the plastic coating of optical fibers.

7.4.1 Enhancement of Reactions in Organic Synthesis by MDEL Systems

Ever since the 1968 studies of Ward and Wishnok [23] on the use of MDELs, such electrodeless lamps have failed for several years to become the light sources of choice in photochemical organic syntheses. Only in the last two decades have they been used for such applications [24–28]. The combined chemical activation by the use of two different types of electromagnetic radiation (viz., microwave and UV/Vis radiations) now covers the field that has become broadly referred to as microwave-assisted photochemical organic synthesis (MAPOS). Synergistic effects of MDELs have been demonstrated on comparison of synthetic methods with light alone and/or with microwave radiation alone [24]. An MDEL apparatus used in organic synthesis (Fig. 7.7) has also been examined in some detail [26, 27].

Additionally, the concentration and nature of the filling compound (gas) in MDELs has been optimized for photochemical reactions [28] and the MDEL light sources have been evaluated in various organic solvents [25–29]. In this regard, Klán and Církv have described some of the features of MDELs and related devices for use in organic syntheses [30].

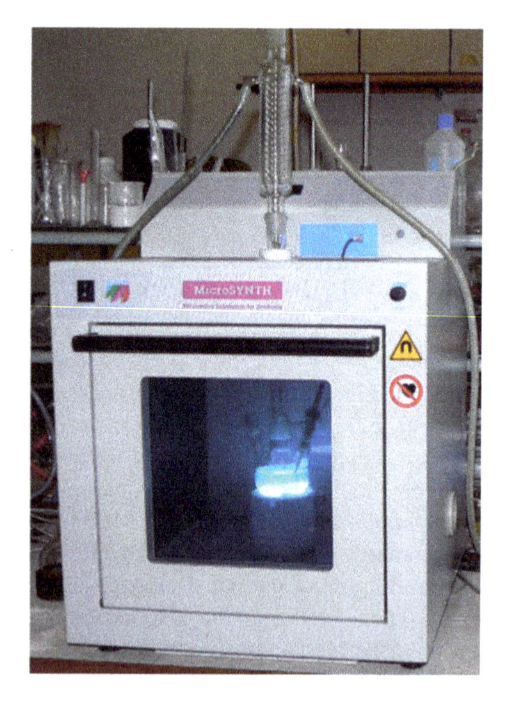

Fig. 7.7 Testing of a MEDL's performance on Milestone's MicroSYNTH Lab Station. Figure and related text reproduced with permission from Církva and coworkers, Ref. [27], via License Number 4714391287566; Copyright 2005 by Elsevier B.V.

7.4.2 Environmental Remediation Using Microwave Dielectric Heating

A principal environmental process to remediate aqueous ecosystems uses microwave radiation in the sterilization of bacterial-contaminated sites. However, since the wavelength and the photon energy of the 2.45 GHz microwaves are only about 12.4 cm and 1×10^{-5} eV, respectively, the microwave energy is about five orders of magnitude smaller than the vibrational energy of typical molecules. Consequently, the energy to cleave a chemical bond is not available from microwaves, and a direct degradation of organic pollutants by microwave irradiation alone is not plausible. Nonetheless, the microwave sterilization technology is a major method in the food industry. By comparison, cases of microwave sterilization of wastewaters are rather scarce.

In recent years, MDELs have proven particularly useful as light sources in the degradation of organic substrates and in environmental remediation. MDELs that generate short wavelength UV light have been examined as possible new sterilization lamps [31]. Device design optimization has been examined in detail. Studies on the use of MDEL light sources in sterilization reactors have been reported [32]. Model wastewater treatments by MDEL systems to degrade Acid Orange 7 and to oxidize various aromatic species in the presence of H_2O_2 have been carried out using small-scale reactors with MDELs emitting UV wavelengths longer than 228 nm [33]. Large-scale sterilization of drinking water (10 L) infected with various microorganisms (e.g., *Escherichia coli*) has also been examined using a microwave stimulated electrodeless lamp emitting UV wavelengths longer than 254 nm [34]. The use of MDELs to sterilize *E. coli* bacteria by generating reactive oxygen species (e.g., ·OH) has been described [35]. Wastewater treatments with MDELs in a batch reactor system [36] and in a flow-through reactor system [34, 35, 37] have been examined on the basis of application scales. However, neither sterilization nor degradation of organic substrates by microwaves alone is expected in many cases. Accordingly, the use of new UV light sources is a rather attractive prospect. In this regard, Fassler and coworkers [38] proposed a circulating wastewater treatment system that combines a microwave domestic oven incorporating a MDEL device (Fig. 7.8). An improved scaled-up device of this type has been developed by UV-EL GmbH & Co. KG [39].

7.5 Advantages of Using a Semiconductor Microwave Generator

We now examine some of the advantages of using a semiconductor microwave generator in generating microwaves and then discuss what some of the benefits in using MDELs might be. The advantage of MDELs over existing lamps is ***fast start-up time and flicker-free arc attachment***. Indeed, a change from a magnetron microwave generator to a semiconductor microwave generator provides supplementary significant benefits.

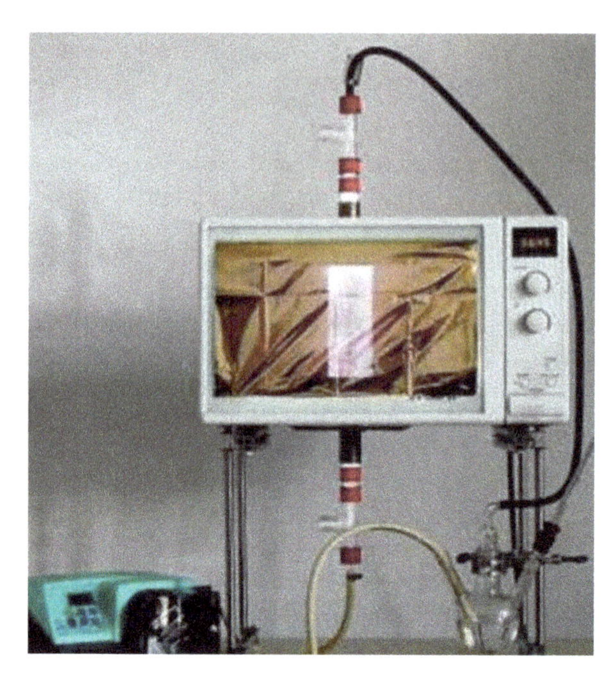

Fig. 7.8 Treatment of wastewater with a double tube UV reactor with microwave excitation (MEDL's performance) in a domestic microwave oven. Reproduced from Ref. [38]. Copyright 2004 by IUVA, P. O. Box 1110, Ayr (ON), Canada NOB 1EO

A magnetron-based generator is generally used in commercial microwave apparatuses, regardless of whether it operates under an industrial-based or a consumer-based protocol. Significant different features of semiconductor and magnetron generators are worth noting. For instance, the frequency distribution of the microwave radiation from the magnetron generator in domestic microwave cooking ovens is shown in Fig. 7.9, which shows that the frequency of the microwaves is distributed over a large frequency range: from about 2.25 to 2.60 GHz [40]. Moreover, the distribution of the microwave frequency changes depending on the characteristics of the microwave generator equipment, the applicator, and the load. Note that narrower-band frequency waves can, in principle, be generated with high accuracy with a magnetron. By contrast, the semiconductor generator produces microwaves within a very narrow frequency range of 2.45000 ± 0.00250 GHz (Fig. 7.9).

In the case of the magnetron generator, the microwave output power distribution indicates a widely dispersed frequency distribution. Therefore, the output of the actual 2.45 GHz microwaves is smaller than the output power. On the other hand, when using a semiconductor microwave generator, microwave heating can evolve efficiently because the microwave output power is concentrated at the 2.45000 GHz frequency. In other words, semiconductor oscillators can transmit microwaves at a very precise frequency, which also means that precise microwave transmission can be adjusted.

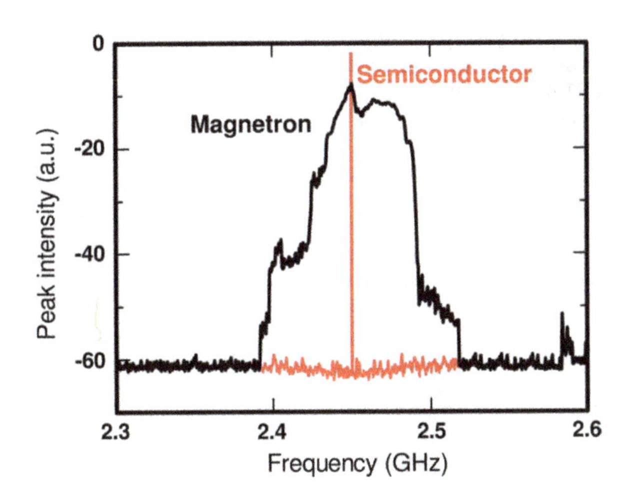

Fig. 7.9 Frequency spectral distribution of the 2.45 GHz microwave radiation emitted from the magnetron generator (black curve) and the semiconductor microwave generator (red curve). Figure and related text reproduced from Horikoshi and coworkers, Ref. [40]. Copyright 2011 by the American Chemical Society

In order to light a MDEL lamp with microwaves, it is necessary to match the impedance of 50 Ω. Furthermore, since the impedance of a non-light-emitting MDEL and the impedance of a light-emitting MDEL are different from each other, tuning is necessary to derive optimum conditions. Normally, matching is performed using a mechanical tuner. However, this tends to increase the overall size of the device. If a semiconductor microwave generator were used, then the frequency could be varied so that a mechanical tuner would not be necessary, in addition to which downsizing of the device and reduction of costs of the device can be achieved. Generally speaking, semiconductor microwave generators tend to be more expensive than magnetron microwave oscillators, but the cost of incidental equipment can be reduced that may reduce the total cost of the equipment. Miniaturization of microwave oscillators enables connection as a unit to various process instruments. For example, Tokyo Keiki Inc. and Plasma Applications Co. Ltd. have reported a prototype water sterilizer that combines a semiconductor microwave oscillator and a MDEL light source (Fig. 7.10) [41]. UVC radiation generated from the MDEL can sterilize water contaminated with *E. coli* in PET bottles and can sterilize up to 1.5–2.0 L of contaminated water within a few seconds.

7.6 Initiatives that a MDEL Lamp Should Aim for in the Future

The Minamata Convention on Mercury, an International Treaty that aimed at regulating the manufacture, import and export of products containing mercury, was established in 2013 and enacted in 2017 [42]. This Treaty requires nearly all countries to regulate the manufacturing and sales of mercury lamps by 2020. Hence, present-day commercially available UV lamps that contain mercury as an inner filler

Fig. 7.10 Photograph of a device to sterilize water in a PET bottle by UVC irradiation for a discardable water sterilization system. Reproduced from Ref. [41]. Copyright 2016 by the New Energy and Industrial Technology Development Organization (NEDO)

gas will see their usage regulated by enacted mercury regulations. In response to such regulations, there has been a significant shift from mercury-containing fluorescent lamps to light-emitting diode (LED) lamps, which use semiconductors and electro-luminescence to generate light. However, in certain processes, such as treatment of contaminated wastewaters and air purification, LED lamps cannot substitute mercury-free lamps because of the lack of generating significant low-wavelength UV radiation, even though LED lamps have prevailed widely as useful luminescent devices that have displayed lifetimes longer than existing mercury UV lamps, a clear advantage of LEDs. By contrast, electroluminescent (EL) lamps or so-called ***high-field electroluminescent*** lamps use electric current directly through a phosphor to generate light; electro-luminescence is the non-thermal conversion of electrical energy into light energy [43].

In developing Hg-free light sources, Horikoshi and Serpone [44, 45] have examined and reported on substituting mercury with another filler gas, while using the electroluminescent (EL) platform. They reported on the optimization of internal sealed filler gas(es) using the barrier discharge powered by an AC power supply. Switching from a microwave device to an AC power source greatly simplified the device and increased the lighting efficiency of these EL light sources. In the earlier study [44], these authors developed a mercury-free MDEL and investigated optimal MDELs that incorporated nitrogen, oxygen, hydrogen, xenon, argon, and helium as the filler gases alone, or as mixed filler gases. The most suitable light plasma with emitted lines concentrated in the 300–400 nm spectral range (under the conditions used) was obtained with a nitrogen/argon filler mixture in a ratio of 20/80 v/v% and at a pressure around 700 Pa.

The principal goal of Horikoshi and coworkers [44, 45] was to mix similar inexpensive and innocuous gases to develop Hg-free EL lamps that generate ultraviolet radiation. As an example, nitrogen, argon, and helium were the base filler gases, while the SF_6 gas (sulfur hexafluoride gas) was the additive filler gas in smaller quantities. Sulfur hexafluoride is a chemically stable, non-toxic, odorless, colorless, and incombustible gas [46], which is widely used in industry as a plasma-etching product that effectively emits ultraviolet radiation [47]. To the extent that the use of Hg as a filler gas was shown earlier [21] to enhance the efficiency of emitting ultraviolet radiation, they substituted SF_6 gas for Hg as the additive filler gas.

Typical photographs and displays of UV-visible spectral lines of the light plasma generated from pure N_2, He, Ar, and SF_6 gases are reported in Fig. 7.11 under optimal pressure conditions selected so as to have the highest UV light intensity at wavelengths below 400 nm. The optimal pressures were: 2500 Pa for N_2 gas (Fig. 7.11a), 4000 Pa for He gas (Fig. 7.11b), 8500 Pa for Ar gas (Fig. 7.11c), and 100 Pa for SF_6 gas (Fig. 7.11d). The efficiency of filler gases to generate intense UV light varied as: Ar > He > N_2 > SF_6. Intense UV radiation was generated at wavelengths below 300 nm in the case of Ar and He as the filler gases.

Moreover, additions of small quantities of SF_6 gas to either He or Ar as the base gases caused significant changes in the UV region as a function of pressure in the range of 10,000–0.1 Pa that was varied in increments of 100 Pa for various mixing ratios. UV radiation was remarkably reduced upon adding the additive SF_6 gas so there was no advantage in mixing SF_6 with either He or Ar. On the other hand, for N_2 as the base gas, the UV generated in the 200 to 300 nm region was rather small.

7.7 Concluding Remarks

It took over a century for the first commercial RF lamp to appear in the case of radio frequency (RF) lighting, which introduced a new era in the generation of light. Recent progress in semiconductor power switching electronics, along with a more thorough understanding of fundamental processes in RF plasmas, have resulted in commercially available light sources [48]. RF light sources follow the same basic principles of converting electrical power into light radiation as conventional gas discharge lamps do. The fundamental difference between RF lamps and conventional lamps is that RF lamps operate without electrodes (i.e., no anode and no cathode). This has profound consequences on the RF lamp's characteristics and features. The elimination of electrodes opens up great opportunities for increased durability, light output, and efficiency. Moreover, it removes many of the restrictions associated with lamp shapes of conventional electrode discharge lamps. The first commercial RF lamps have been directed mainly toward niche applications where high costs of installation and maintenance offset the rather large initial costs of the lamps. The initial cost of RF lighting products is the major barrier to the widespread use of

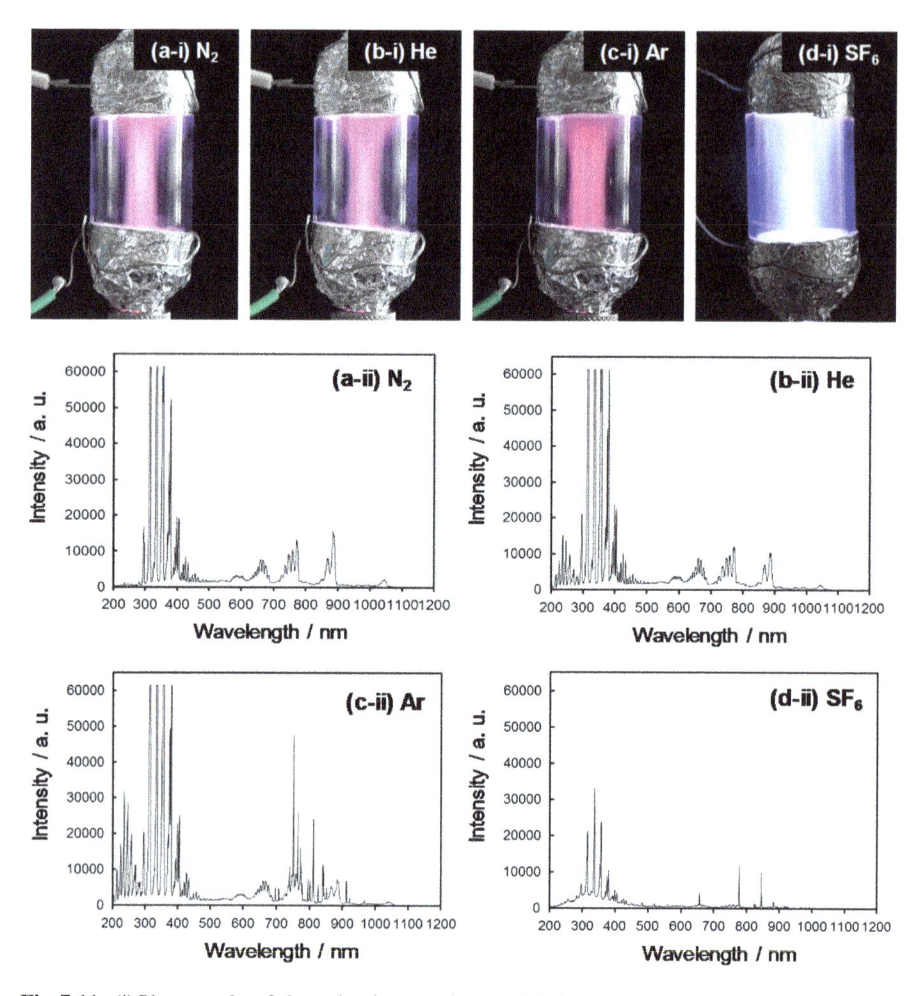

Fig. 7.11 (i) Photographs of electroluminescent lamps with single filler gases and (ii) the UV/Vis/IR emission spectral lines from the electroluminescent lamps; the filler gases were: **a** N_2, **b** He, **c** Ar and **d** SF_6. Note that, as shown in the actual photographs of the EL lamps in (a–i to d–i), the electrodes consisted of an aluminum foil externally wrapped around the extremes of the bulb ampoule, with the bare AC electrical wires wrapped around the aluminum foil. Accordingly, the distance between the electrodes is taken as the distance between the external aluminum foil. Figure and related text reproduced from Horikoshi and coworkers, Ref [45]; Copyright by the European Society for Photobiology, the European Photochemistry Association, and The Royal Society of Chemistry

RF lamps. However, with further advances in the use of semiconductor microwave generators in RF lighting technology, the range of applications will no doubt witness significant increases.

References

1. http://www.edisontechcenter.org/incandescent.html#inventors. Accessed Oct 2019
2. Edison TA (1880) Electric lamp. US Patent 223,898, 1880
3. Edison TA (1907) Fluorescent electric lamp. US Patent 865,367, 1907
4. http://www.edisontechcenter.org/flourescent.html#inventors. Accessed Oct 2019
5. Bright AAJ (1949) The electric-lamp industry. MacMillan Publishers, New York
6. http://edisontechcenter.org/DFMoore.html. Accessed Oct 2019
7. https://www.americanhistory.si.edu/lighting/bios/moore.htm. Accessed Oct 2019
8. http://www.lamptech.co.uk/Documents/M6%20Cooper%20Hewitt.htm; http://www.lamptech.co.uk/Documents/M6A%20Lamp%20Design.htm. Accessed Oct 2019
9. See for example: http://www.lamptech.co.uk/Spec%20Sheets/D%20ED%20Osram%20Endura.htm. Accessed Oct 2019
10. http://www.neonlibrary.com/neon_history_2.html. Accessed Oct 2019
11. Hewitt PC (1907) Method of producing electric light. US Patent 843,534, 1907
12. Hewitt PC (1910) Induction lamp. US Patent 966,204, 1910
13. Bouman et al (1987) Electrodeless low-pressure gas discharge lamp. US Patent 4,645,967, 1987
14. https://en.wikipedia.org/wiki/Electrodeless_lamp. Accessed Oct 2019
15. Godyak VA (1998) High intensity electrodeless low pressure light source driven by a transformer core arrangement. US Patent 5,834,905, 1998
16. http://www.plasma-i.com/plasma-history.htm. Accessed Oct 2019
17. Johnston CW, van der Heijden HWP, Janssen GM, van Dijk J, van der Mullen JJAM (2002) A self-consistent LTE model of a microwave-driven, high-pressure sulfur lamp. J Phys D Appl Phys 35:342–351
18. Turner BP, Ury MG, Leng Y, Love WG (1997) Sulfur lamps—progress in their development. J Illuminat Eng Soc 26:10–16
19. Ciolkosz DE, Albright LD, Sager JC (1998) Microwave lamp characterization. Life Support Biosph Sci 5:167–174
20. Krizek DT, Mirecki RM, Britz SJ, Harris WG, Thimijan RW (1998) Spectral properties of microwave-powered sulfur lamps in comparison to sunlight and high pressure sodium/metal halide lamps. Biotronics 27:69–80
21. Horikoshi S, Abe M, Serpone N (2009) Novel designs of microwave discharge electrodeless lamps (MDEL) in photochemical applications. Use in advanced oxidation processes. Photochem Photobiol Sci 8:1087–1104
22. Ashikaga K, Kawamura K (2006) UV irradiation equipment using DC power supply. Kougyo Toso 201:40–44 (Japanese)
23. Ward HR, Wishnok JS (1968) The vacuum ultraviolet photolysis of benzene. J Am Chem Soc 90:5353–5357
24. Církva V, Hájek M (1999) J Photochem Photobiol A: Chem 123:21–23
25. Klán P, Literák J, Hájek M (1999) The electrodeless discharge lamp: a prospective tool for photochemistry. J Photochem Photobiol A: Chem 128:145–149
26. Klán P, Hájek M, Církva V (2001) The electrodeless discharge lamp: a prospective tool for photochemistry Part 3. The microwave photochemistry reactor. J Photochem Photobiol A: Chem 140:185–189
27. Církva V, Vlková L, Relich S, Hájek M (2006) Microwave photochemistry IV: preparation of the electrodeless discharge lamps for photochemical applications. J Photochem Photobiol A: Chem 179:229–233
28. Müllera P, Klán P, Církv V (2005) The electrodeless discharge lamp: a prospective tool for photochemistry Part 5: fill material-dependent emission characteristics. J Photochem Photobiol A: Chem 171:51–57
29. Církv V, Kurfürstová J, Karban J, Hájek M (2004) Microwave photochemistry II. Photochemistry of 2-tert-butylphenol. J Photochem Photobiol A: Chem 168:197–204

30. Klán P, Církva V (2006) Microwaves in organic synthesis. In: Loupy A (ed) Wiley–VCH Verlag, Weinheim, Germany, pp 860–897. Chapter 19
31. Al-Shamma'a AI, Pandithas I, Lucas J (2001) Low-pressure microwave plasma ultraviolet lamp for water purification and ozone applications. J Phys D Appl Phys 34:2775–2781
32. Howard AG, Labonne L, Rousay E (2001) Microwave driven ultraviolet photo-decomposition of organophosphate species. Analyst 126:141–143
33. Klán P, Vavrik M (2006) Non-catalytic remediation of aqueous solutions by microwave-assisted photolysis in the presence of H_2O_2. J Photochem Photobiol A: Chem 177:24–33
34. Bergmanna H, Iourtchouk T, Schöps K, Bouzek K (2002) New UV irradiation and direct electrolysis—promising methods for water disinfection. Chem Eng J 85:111–117
35. Iwaguch S, Matsumura K, Tokuoka Y, Wakui S, Kawashima N (2002) Sterilization system using microwave and UV light. Coll Surf B: Biointerfaces 25:299–304
36. Florian D, Knapp G (2001) Anal Chem 73:1515–1520
37. http://www.umex.de. Accessed 2019
38. Fassler D, Drewitz A, Thomas Ch, Meyer A, Johne St. (2003) Proceedings 9th international conference AOTs-9, Montreal (Quebec), Canada, Oct 2003
39. http://www.uv-el.de/uv_el-d.de/index.htm. Accessed Oct 2019
40. Horikoshi S, Osawa A, Abe M, Serpone N (2011) On the generation of hot-spots by microwave electric and magnetic fields and their impact on a microwave-assisted heterogeneous reaction in the presence of metallic Pd nanoparticles on an activated carbon support. J Phys Chem C 115:23030–23035
41. Feasibility study on research and development of UV-C generators using compact, high-efficiency GaN oscillators. New Energy and Industrial Technology Development Organization (NEDO) project, 2015–2016, Japan
42. http://www.mercuryconvention.org/. Accessed Sept 2019
43. http://www.edisontechcenter.org/electroluminescent.html. Accessed Sept 2019
44. Horikoshi S, Kajitani M, Sato S, Serpone N (2007) A novel environmental risk-free microwave discharge electrodeless lamp (MDEL) in advanced oxidation processes, degradation of the 2,4-D herbicide. J Photochem Photobiol A: Chem 189:355–363
45. Horikoshi S, Yamamoto D, Hagiwara K, Tsuchida A, Matsumoto I, Nishiura Y, Kiyoshima Y, Serpone N (2019) Development of a Hg-free UV light source and its performance in photolytic and photocatalytic applications. Photochem Photobiol Sci 18:328–335
46. Hughes TG, Smith RB, Kiely DH (1983) Stored chemical energy propulsion system for underwater applications. J Energy 7:128–133
47. Sansonetti JE, Martin WC (2005) Handbook of basic atomic spectroscopic data. J Phys Chem Ref Data 34:1559–2259
48. Godyak VA (2002) Bright idea: radio-frequency light sources. IEEE Ind Appl Mag 8:42–49

Part III
Energy Applications

Chapter 8
Microwave Plasma

Hirotaka Toyoda

Abstract Microwave plasma is one of the important plasma sources in industry because of its high-plasma density and low damage to the processing surface. In this chapter, we examine the basic physics of the production of microwave plasma from the discharge breakdown to plasma sustainment, together with various production techniques of microwave plasma from low-pressure surface wave plasma, electron cyclotron plasma to atmospheric pressure microwave plasmas. Recently, a solid-state-based microwave power source has been utilized as the power source of microwave plasmas as an alternative to conventional magnetron-type microwave power supplies. Benefits of the solid-state microwave power for the plasma source are explained by focusing on the controllability of the microwave phase of the solid-state microwave power source.

8.1 Microwave Discharge Breakdown

In microwave plasma generation, microwave power is used at a typical frequency of 2.45 GHz for the generation of plasma; the frequency is much higher than those of typical RF plasmas (13.56 MHz). From the viewpoint of production and loss of charged particles, however, discharge breakdown by microwaves can be considered the same as those of other discharges such as DC or RF, with breakdown occurring when the rate of ionization exceeds the rate of loss of charged particles. Accordingly, let us consider an electron accelerated by the microwave's electric field. Electron speed can be obtained from the following equation.

$$m\frac{\mathrm{d}v}{\mathrm{d}t} = -eE - m\nu_{\mathrm{c}}v \tag{8.1}$$

where m and e are the electron mass and electron charge, respectively; ν_{c} is the electron collision frequency, which is proportional to the neutral species density n_g.

H. Toyoda (✉)
Department of Electronics, Nagoya University Furo-cho, Chikusa-ku, Nagoya 464-8603, Japan
e-mail: toyoda@nuee.nagoya-u.ac.jp

S. Horikoshi and N. Serpone (eds.), *RF Power Semiconductor Generator Application in Heating and Energy Utilization*, https://doi.org/10.1007/978-981-15-3548-2_8

Supposing sinusoidal oscillation of the electric field at radian frequency of ω as given by Eq. 8.2,

$$E = E_0 \sin(\omega t) \tag{8.2}$$

then the electron oscillates at a speed given by Eq. 8.3:

$$v = \frac{eE_0}{m\sqrt{\omega^2 + \nu_c^2}} \sin(\omega t - \phi) \tag{8.3}$$

with a phase difference of $\phi = \mathrm{tain}^{-1}(\omega/\nu_c)$. From the phase difference between the electric field and the electron speed, the average power obtained by the electron becomes (Eq. 8.4),

$$W = \frac{(eE_{\mathrm{eff}})^2}{m\nu_c} \tag{8.4}$$

where

$$E_{\mathrm{eff}} = \sqrt{\frac{\nu_c^2}{\nu_c^2 + \omega^2}} E_0 \tag{8.5}$$

The electron energy W becomes its maximum when $\omega = \nu_c$, which corresponds to ca. 1 kPa at a microwave frequency of 2.45 GHz. The above discussion is for the case of a single electron colliding with neutral species. If the initial electron gained enough kinetic energy to ionize neutral species, second and further generation of electrons are produced in the gas phase by the ionization of neutral species. These electrons experience elastic and inelastic collision processes and lead to a certain energy distribution. Although most electrons have much lower kinetic energy than the ionization energy, a small fraction of high energy electrons in the tail of the energy distribution contribute to ionize neutral species. However, the ionization process alone does not determine the condition of breakdown. In parallel with the electron energy gain by the external electric field, electrons diffuse away from the gas phase and are lost on the wall or, in some cases, in the gas phase by ion–electron recombination or electron attachment to neutral species. Neglecting the recombination and electron attachment, particle balance of electrons is described by Eq. 8.6:

$$\frac{\mathrm{d}n}{\mathrm{d}t} = \nu_i n + D\nabla^2 n \tag{8.6}$$

where n is the electron density, ν_i is the ionization frequency determined by the ionization rate coefficient $k_{iz,}$ and the density of neutral species n_g as $\nu_i = k_{iz}\, n_g$. D denotes the diffusion constant of electrons in the gas phase. Considering a vacuum vessel of a characteristic size Λ, discharge breakdown occurs when the ionization

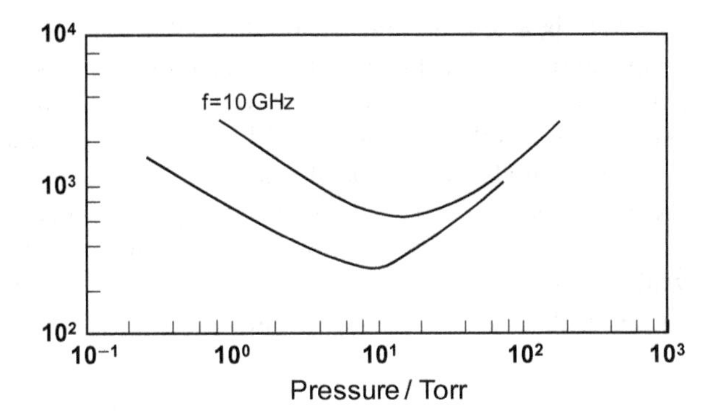

Fig. 8.1 Breakdown microwave electric field as a function of the pressure. Microwave frequencies are 3 and 10 GHz. Reproduced from Ref. [1]. Copyright 1991 by Springer, Berlin

rate exceeds the rate of electron loss by diffusion to the wall. The following condition holds as the breakdown condition (Eq. 8.7),

$$\nu_{\mathrm{i}} > \frac{D}{\Lambda^2} = \frac{1}{\tau_D} \tag{8.7}$$

where τ_D is the typical lifetime of charged particles diffusing to the wall. Please note that the diffusion constant is inversely proportional to the density of neutral species in the gas phase. Considering that ν_{i} is proportional to n_g and that the electron energy increases with n_g if $\omega > \nu_{\mathrm{c}}$, then the breakdown becomes easier with increasing n_g at $\omega > \nu_{\mathrm{c}}$. However, when the n_g is further increased, the electron energy decreases as shown in Eq. 8.7 and ionization then becomes difficult. Figure 8.1 shows the measured breakdown electric field as a function of gas pressure [1]. The solid curve shows the calculated result including the energy gain of electrons by the microwaves' electric field and the energy distribution of electrons through collision processes. Electric field breakdown varies strongly, depending on the pressure and shows a minimum at ~1 Pa, showing the optimum neutral density for the microwave breakdown.

8.2 Establishment of Steady-State Discharge

Although microwave breakdown was discussed in the above section, this condition is not always the case of plasma ignition. To obtain steady-state plasma, ionization of neutral species in the plasma and ion–electron loss from the plasma must be balanced [2]. Let us then consider the balance of charged particles in the plasma with a zero-dimensional model. In this model, we assume that the plasma density n and the electron temperature T_{e} are uniform in the vessel, which is often the case of low-pressure discharges. Supposing a plasma volume V, the total number of ion–electron pairs G produced in the plasma at a unit time interval is expressed by Eq. 8.8 as

$$G = k_{\mathrm{iz}}(T_{\mathrm{e}}) n_g n V \tag{8.8}$$

where k_{iz} is again the ionization rate coefficient and is a function of the electron temperature T_e; n_g is the density of neutral species. At the same time, ion–electron pairs are lost from the plasma to the wall through a sheath formed between the plasma and the wall. In the case of typical low-pressure plasmas, the electron temperature T_e is much higher than that of ions (T_i). The sheath is formed naturally because the plasma has to keep its charge neutrality. The sheath forms a potential difference between the plasma and the wall, with the electrons tending to diffuse away to the wall with higher kinetic energies are reflected by the potential barrier formed by the sheath. At the edge of the sheath, the flux of ions Γ_i is given by Eq. 8.9,

$$\Gamma_i = n_s\sqrt{\frac{kT_e}{M}} \tag{8.9}$$

where n_s and M represent the plasma density at the sheath edge and ion mass, respectively. In a simple model, n_s is related to the bulk plasma density n as Eq. 8.10,

$$n_s = e^{1/2}n \tag{8.10}$$

where e is the base of the natural logarithm. Ions are accelerated and electrons are decelerated in the sheath and ion and electron fluxes at the wall become the same. Assuming a surface area of the plasma boundary to be S, the total loss of charged particles (L) from the plasma at a unit time interval is (Eq. 8.11),

$$L = e^{1/2}n\sqrt{\frac{kT_e}{M}}S \tag{8.11}$$

The balance equation of charged particles in the plasma then becomes (Eq. 8.12).

$$k_{iz}(T_e)n_g nV = e^{1/2}n\sqrt{\frac{kT_e}{M}}S \tag{8.12}$$

The latter equation shows that the balance is not determined by the plasma density, but is determined by the ionization rate coefficient, the electron temperature, the plasma volume, and the plasma surface area. As k_{iz} is positive (and strong), the function T_e is an important parameter to establish steady-state plasma.

8.3 Electromagnetic Wave Propagation in Plasma

As charged particles such as electrons or ions are mobile in the space, these species are influenced by the electromagnetic wave propagating in the plasma and moves with the same frequency as that of the electromagnetic wave (ω). Let us consider a condition where electrons and ions are mobile and stationary, respectively. This condition stands because the ion mass is much larger than that of electron. Furthermore, we neglect the electron collision process, supposing that wave frequency ω is higher than collision frequency ν. In this case, the plasma behaves as a dielectric matter with varying permittivity depending on the plasma density. Relative permittivity of the plasma (ε_p) is expressed as Eq. 8.13.

$$\varepsilon_p = 1 - \left(\frac{\omega_p}{\omega}\right)^2 \quad (8.13)$$

where ω_p is taken as the plasma (oscillation) frequency and is described by Eq. 8.14,

$$\omega_p = \sqrt{\frac{e^2 n}{\varepsilon_0 m}} \quad (8.14)$$

where n, m, and ε_0 are the plasma density, the electron mass, and the vacuum permittivity, respectively. Figure 8.2 shows the relative plasma permittivity as a function of wave frequency normalized by the plasma frequency. The interesting feature of the plasma permittivity is that the relative permittivity is always less than unity and even negative permittivity is possible depending on the relation between the plasma density and the wave frequency. If the wave frequencies ω were larger than the plasma frequency ω_p, then the electromagnetic waves would propagate through the plasma. However, if the wave frequency came across a critical condition of $\omega_p = \omega$, and became lower than the plasma frequency ω_p, then the wave could not exist in the plasma. In this situation, the plasma becomes a kind of metal, causing the waves injected into the plasma from the outside to be reflected.

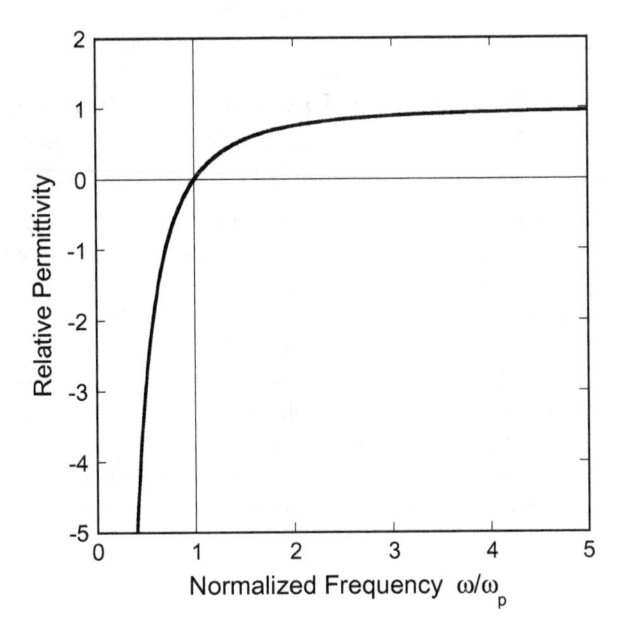

Fig. 8.2 Frequency dependence of relative plasma permittivity

Fig. 8.3 Plasma density dependence of plasma frequency. Electromagnetic wave is reflected when the wave frequency is lower than the plasma frequency ω_p

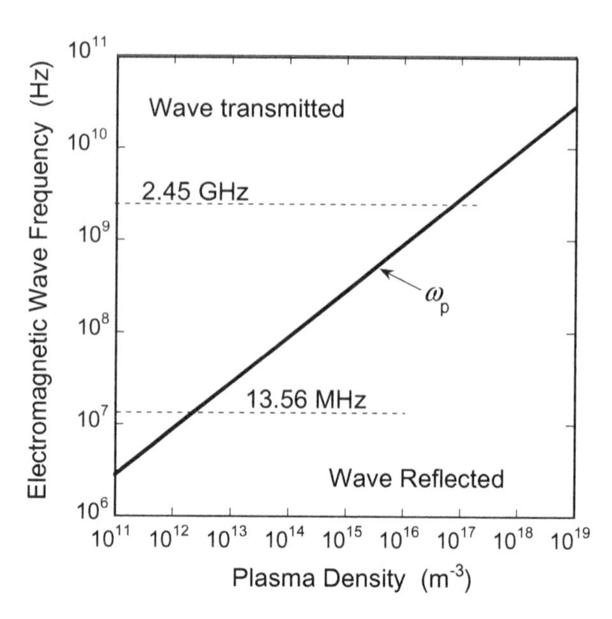

Figure 8.3 shows schematically the boundary of wave reflection and wave propagation in a two-dimensional space of the plasma density and the electromagnetic wave frequency. At a frequency of 13.56 MHz—the conventional RF frequency used for plasma production—the electromagnetic wave is reflected at very low plasma density of 2.7×10^{12} m^{-3}. At a microwave frequency of 2.45 GHz, the critical density of the wave reflection becomes very high (7.4×10^{16} m^{-3}).

8.4 Production of Low-Pressure Microwave Plasma Without Magnetic Field (Surface Wave Plasma; SWP)

The influence of the collision process to wave propagation in the plasma becomes rather small at moderately low pressures and when the collision frequency is lower than the wave frequency. In the case of 2.45 GHz microwaves, the collisionless condition stands at pressures less than ~10 Pa, because the collision frequency ν is less than ~10^9 s^{-1} and is less than the radian frequency of the 2.45 GHz microwaves (1.54×10^{10} s^{-1}). When the plasma is produced by microwaves and the microwave power is introduced to the plasma through a dielectric plate, the surface wave can be excited and the plasma is then sustained with the aid of the surface wave power distribution [3–7]. The dispersion relationship, namely the relationship between the frequency ω and the wavenumber k, of the surface wave is given by Eq. 8.15:

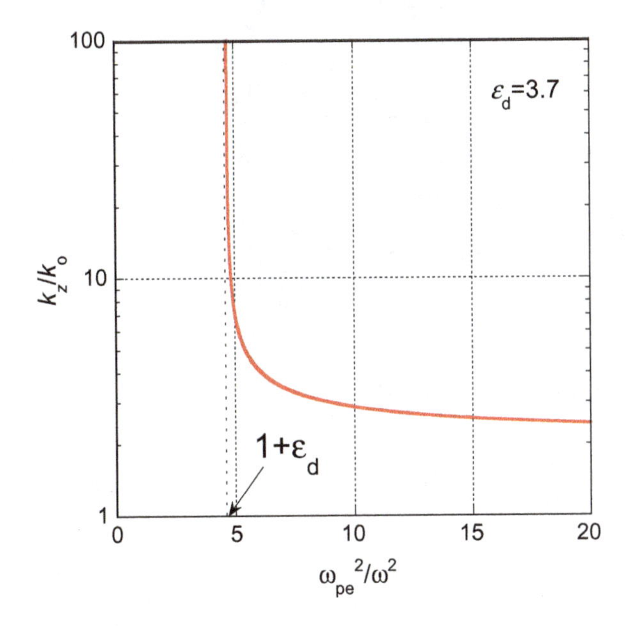

Fig. 8.4 Dispersion relation of surface wave

$$k = \frac{\omega}{c}\sqrt{\frac{\varepsilon_{\mathrm{d}}\left(\omega_{\mathrm{p}}^2 - \omega^2\right)}{\omega_{\mathrm{p}}^2 - (1 + \varepsilon_{\mathrm{d}})\omega^2}} \tag{8.15}$$

where ω_{p} and ε_{d} denote the plasma frequency and the dielectric permittivity, respectively. Figure 8.4 shows the relationship between the wavenumber and the square of the plasma frequency (corresponding to the plasma density from Eq. 8.14). The figure indicates that the surface wave does not exist when the ratio $\omega_{\mathrm{p}}^2/\omega^2$ is less than $1 + \varepsilon_{\mathrm{d}}$. Whenever $\omega_{\mathrm{p}}^2/\omega^2 = 1 + \varepsilon_{\mathrm{d}}$, the frequency ω is then referred to as a surface wave resonant frequency ω_{sw} and is determined from Eq. 8.16:

$$\omega_{\mathrm{sw}} = \frac{1}{\sqrt{1 + \varepsilon_{\mathrm{d}}}}\omega_{\mathrm{p}} \tag{8.16}$$

When the plasma density becomes high and the surface wave resonance frequency ω_{sw} exceeds the microwave frequency ω, the microwave power begins to propagate along the interface between the plasma and the dielectric. Considering 2.45 GHz microwaves and a quartz plate ($\varepsilon_{\mathrm{d}} = 3.7$) as a dielectric plate, the minimum plasma density required to excite the surface wave obtained from Eq. 8.16 is 3.6×10^{17} m^{-3}. An interesting feature of the SWP is the possibility of plasma elongation along the dielectric surface with the aid of the microwave power distribution with the surface wave excited at the dielectric–plasma boundary. One of the typical SWP configurations is shown in Fig. 8.5. A quartz tube goes through a rectangular waveguide and the SWP is produced inside the tube. The microwave power propagates along the quartz tube and one-dimensionally long SWP is produced along the quartz tube.

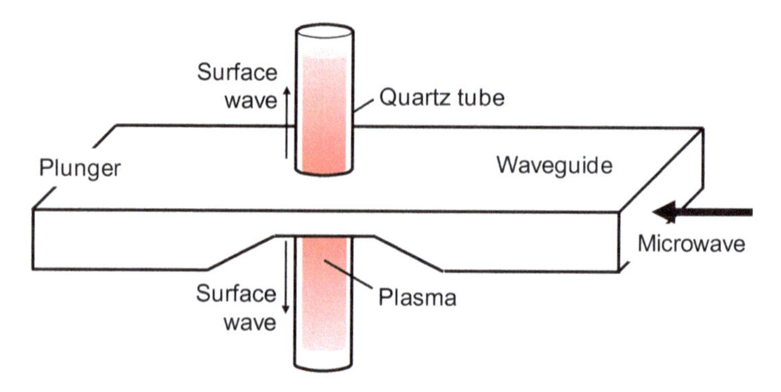

Fig. 8.5 Example of surface wave plasma production in dielectric tube

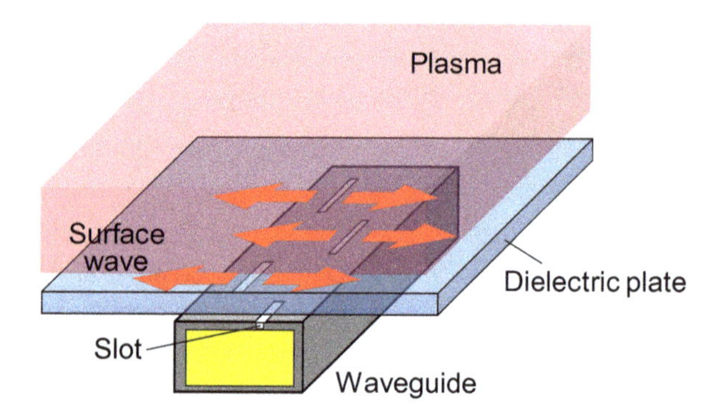

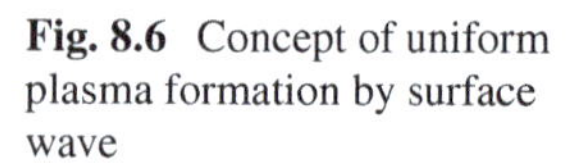
Fig. 8.6 Concept of uniform plasma formation by surface wave

Another SWP is the planar surface wave plasma source [8–15]; its configuration is schematically illustrated in Fig. 8.6. In this case, the microwave power is radiated through slots placed in a rectangular waveguide. Slots are placed at a distance half the wavelength in the waveguide. The slot length is optimized so as to maximize the microwave power radiation from slots. A dielectric plate is placed adjacent to the slots, and plasma is produced. As in the case of the tubular SWP, the microwave power propagates along the interface of the dielectric and the plasma. Typically, a TM-mode surface wave propagates along the interface, and the electromagnetic power penetrates the plasma as a momentary wave. The electric field perpendicular to the dielectric surface accelerates the electrons in the plasma, especially at a certain plasma density where the plasma frequency coincides with the microwave frequency. Accelerated electrons are injected into the plasma to promote ionization in the plasma. In this configuration, two-dimensionally large SWP can be produced and a SWP with a size of a few square meters has been reported. In planar surface wave plasma, the discharge mode is sometimes discussed. The dielectric plate is surrounded by a metal vessel and, when a surface wave is excited on a dielectric surface, various modes corresponding to the geometry of the dielectric surface become possible and the standing wave on the dielectric surface produces a mode pattern of the plasma. On the other hand, the wave number of the surface wave is dependent on the plasma density,

Table 8.1 Comparison of magnetron and solid state as power supply of microwave plasma sources

Type of power source	Magnetron	Solid state
Frequency	Typically 2.45 GHz	Higher than 2.45 GHz
Frequency stability	Unstable	Stable
Phase control	Impossible	Possible
Power output	kW or higher	~1 kW
Cost/power	Low cost	High depending on power
Size of power source	Large space required	Compact

as shown in Fig. 8.4, and the mode pattern varies depending on the wavenumber, i.e., the plasma density. When the microwave power is varied, a mode jump of the plasma, i.e., change of the mode pattern, is observed. The SWP has been applied to plasma CVD and etching processes [16–19]. The SWP is also applied to polymer surface treatment [20–22] and sputtering processes [23–27].

So far, most of the microwave powers have been produced by magnetron tubes for which the microwave frequency varies depending on the microwave power. However, solid-state microwave generator devices operating at microwave frequency ranges become commercialized and solid-state microwave power generators are available. Comparison of magnetron- and solid-state-type microwave power sources is shown in Table 8.1. Recently, utilizing merits of solid-state microwave power sources, a solid-state microwave plasma sources are developed for better plasma uniformity. Combining the solid-state power amplifier and the signal generator, phase control of the microwave power and plasma control become possible. Using this new technology, a new SWP source using the microwave phase control has been proposed [28, 29]. Figure 8.7 shows a SWP source produced by two solid-state microwave power amplifiers. The microwave signal from a signal generator is divided into two

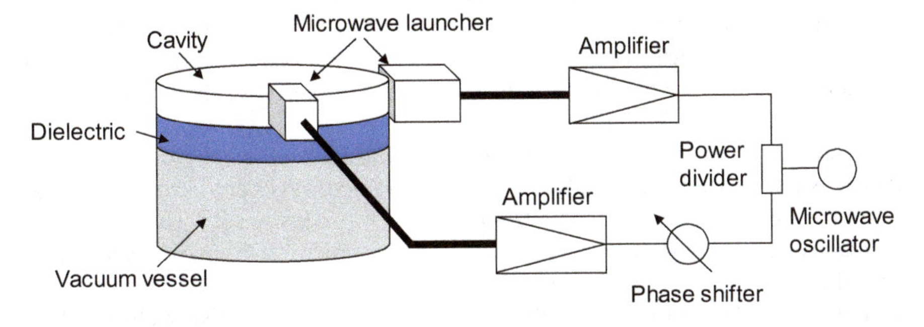

Fig. 8.7 Surface wave plasma source with better uniformity using phase-controlled microwave power supply

lines. The divided signals are amplified by two independent solid-state microwave amplifiers. By inserting a phase shifter in one line, the microwave power from two amplifiers can be phase shifted. Microwave powers are injected into a cylindrical microwave cavity on the top of the vessel from two injection ports that are separated by 90° in an azimuthal position. A dielectric plate at the bottom of the cavity acts as a vacuum window and produces the SWP in the vacuum vessel. Depending on the cavity mode generated in the cavity and the surface wave wavelength, various 2D mode patterns are produced by the surface wave at the interface between the dielectric and the plasma. Generally, the plasma density profile is strongly influenced by the surface wave mode pattern. In the phase-controlled SWP, however, microwave power distribution in the cavity can be varied by the phase shifter. For example, if the phase difference is increased with time, the microwave power distribution rotates with time. By this technique, plasma uniformity on the SWP is controlled and improved.

8.5 Low-Pressure Microwave Plasma with Magnetic Field (Electron Cyclotron Resonance Plasma; ECR Plasma)

By applying magnetic field to the discharge vessel, microwave plasma generation becomes easier and high-density microwave plasma is sustained at very low pressures below 0.1 Pa. In a space with a magnetic field, charged particles gyrate around the line of the magnetic field. Gyration frequency of the electron, i.e., electron cyclotron frequency ω_{ce} can be denoted as Eq. 8.17.

$$\omega_{ce} = \frac{eB}{m} \tag{8.17}$$

where m is again the electron mass and B is the magnetic flux density. When electrons are observed along the direction of the magnetic field line, they gyrate to the right (clockwise direction). A polarized microwave can be considered as a superposition of left- and right-handed circular polarized waves of the same amplitude if the frequency of the microwave coincides with the electron cyclotron frequency. Electrons are constantly accelerated by the right-hand circularly polarized microwave field and the microwave power is easily converted to an electron kinetic energy (electron cyclotron resonance). In the case of 2.45 GHz microwaves, the density of the magnetic field for the resonance is 0.0875 T.

Figure 8.8 shows a schematic of the ECR plasma source. The magnetic flux density is intentionally varied along the axis and microwaves are introduced to the vessel through a dielectric window from the high flux density side. This is because microwaves can propagate only when $\omega_{ce} > \omega$, i.e., microwave can propagate only from the direction of high magnetic field to the resonant magnetic field position. At a position where of $\omega_{ce} = \omega$, the microwave power is effectively absorbed by the electrons in the plasma and high energy electrons then ionize neutral species efficiently even at very low pressures below 0.1 Pa.

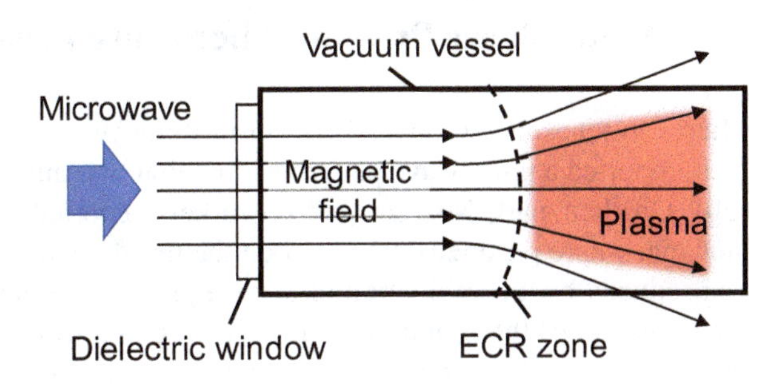

Fig. 8.8 Schematic of ECR plasma source

8.6 High-Pressure Microwave Plasma

As mentioned before, the microwave power is reflected by the plasma when the microwave frequency is lower than the plasma frequency. If the collision frequency is lower than the microwave frequency, the penetration depth (δ) of the microwave power into the plasma is obtained from Eq. 8.18:

$$\delta = \frac{c}{\omega_{\mathrm{p}}} \tag{8.18}$$

where c and ω_{p} are the speed of light in vacuum and the plasma frequency, respectively. This penetration depth is called as collision-less skin depth and is determined by the plasma density and the microwave frequency. However, at high pressures where the collision frequency dominates the microwave frequency, the collision inhibits electron movement and the shielding of the microwave by moving electrons becomes less effective. In this situation, the microwave power tends to penetrate the plasma more deeply compared with the case of the collision-less microwave penetration. When $\nu > \omega$, the equation of the skin depth (i.e., Eq. 8.18) is modified and is expressed as Eq. 8.19.

$$\delta = \left(\frac{c}{\omega_{\mathrm{p}}}\right)\left(\frac{2\nu}{\omega}\right) \tag{8.19}$$

This is referred to as the collisional skin depth and is similar to the conventional skin depth discussed in electromagnetism. This plasma is operated at a rather high-pressure range of ~1 kPa and can produce high-density plasma. Unlike the low-pressure (~10 Pa) surface wave plasma, the microwave electric field penetrates into the bulk plasma and the microwave power directly heats bulk electrons resulting in higher electron temperatures compared with those of surface wave plasmas.

To realize this type of plasma source, the vessel size is sometimes considered. To ease plasma ignition, the vessel size is designed so as to realize cavity resonance in the vessel. To control the cavity resonance, a stage for the processing is installed as a movable one. This type of plasma source is often used for diamond film deposition.

8.7 Atmospheric Pressure Microwave Plasma (APMP)

Microwave plasma is produced even in an atmospheric pressure environment. Moisan et al. reported a microwave plasma source that operates under atmospheric pressure [30]. On their part, Mizojiri and coworkers reported APMP production using an antenna, where plasma is produced on the tip of a wire [31]. Torch type APMP can be produced by a combination of a waveguide and a metal tube [32, 33]. Here, the metal tube goes through a hole of the waveguide from the inside the waveguide to the outside. The tube acts as an antenna and plasma is produced by a strong electric field on the tube tip. Another way to produce the APMP is the use of a discharge gap. In this configuration, the plasma is ignited by a strong electric field produced in the gap between two metal edges. In this type of microwave discharge, various plasma sources have been developed. Hopwood and coworkers reported plasma production based on a microstrip split-ring resonator [34, 35]. A microwave plasma produced in the gap of a microstrip line was also reported [36, 37]. Kono and coworkers reported a microwave discharge in a narrow gap [38, 39].

The above plasma source, however, are rather small in their size and, for their application to large area processing, are not suitable. As one way of producing atmospheric pressure microwave plasma for large area, Toyoda and coworkers have developed a 10 GHz microwave plasma source that is composed of two parallel waveguide lines with an array of 41 slots each, and have succeeded in producing arrayed plasma with an array length of 1 m [40]. This plasma source is still not spatially continuous but is discrete because many small slots are installed as an array in accordance with antinodes of the standing wave inside the short-ended waveguide, and because plasmas in each slot have a length of a few centimeters that is shorter than the distance between adjacent slots. To overcome this issue, the most important point is to suppress the standing wave inside the waveguide. If only a travelling wave inside the waveguide is utilized, the microwave power along the waveguide becomes uniform and, by combining this travelling wave with a very long slot along the waveguide, a uniform and one-dimensionally long atmospheric pressure microwave plasma is expected. Based on this concept, Toyoda et al. have developed a new type of atmospheric pressure microwave plasma source [41–43].

Figure 8.9 schematically shows the plasma source configuration. To realize the travelling wave, a loop-structured waveguide with a microwave circulator is used.

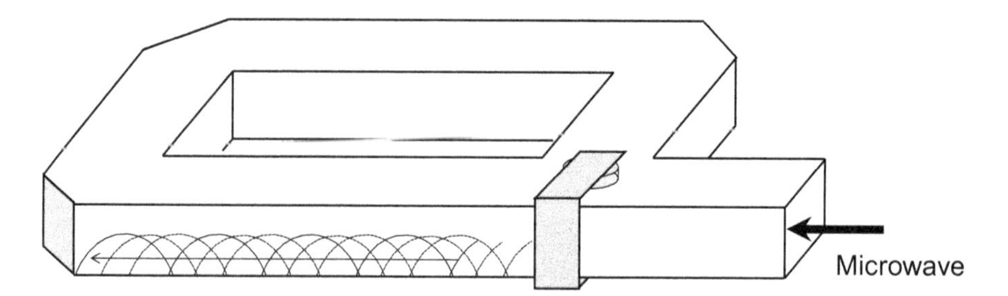

Fig. 8.9 Configuration of travelling microwave formation

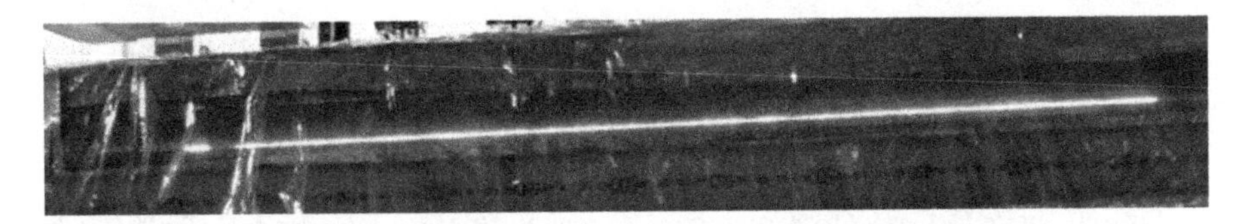

Fig. 8.10 Photograph of plasma emission of AP-MLP

Microwave power from the power source is applied to the looped waveguide through a tuner. The power flow in the looped waveguide is controlled by the circulator and a travelling wave is realized inside the looped waveguide. Microwave power coming back to the power source is reflected again by the tuner and, by the combination of the looped waveguide, the circulator and the tuner, a resonator with the travelling wave is realized.

Figure 8.10 shows an example of plasma emission. Very long atmospheric pressure plasma is produced using Ar. Furthermore, atmospheric pressure plasma using molecular gas is also successfully produced. To realize this, the cross-sectional structure of the waveguide is modified from the conventional rectangular structure to an asymmetric one. By this modification, surface current in the vicinity of the slot is enhanced to support the plasma sustainment with a molecular gas. A nitrogen (N_2) plasma with 50 cm is realized in this plasma source configuration.

References

1. Raizer YP (1991) Gas discharge physics. Springer, Berlin, p 139
2. Liebermann MA, Lichtenberg AJ (1994) Principles of plasma discharges and materials processing. Wiley, New York Chapter 10
3. Glaude MM, Moisan M, Pentel R, Leprince P, Marec J (1980) J Appl Phys 51:5693
4. Moisan M, Shivarova A, Trivelpiece AW (1982) Plasma Phys 24:1331
5. Mateev E, Zhelyazkov I, Atanassov V (1983) J Appl Phys 54:3049
6. Moisan M, Zakrzewski Z (1992) Microwave excited plasmas. In: Moisan M, Pelletier J (eds) Elsevier, Amsterdam, p 123
7. Ferreira CM, Moisan M (1985) Surface waves in plasmas and solids. In: Vukovic S (ed) World Scientific, Singapore, p 113
8. Moisan M, Ferreira CM, Hubert J, Margot J, Zakrzewski Z (1995) Phenomena in ionized gases. In: Becker KH, Carr WE (eds) AIP Press, Woodbury, p 25
9. Ishijima T, Toyoda H, Takanishi Y, Sugai H (2011) Jpn J Appl Phys 50:36002
10. Ghanashev I, Nagatsu M, Sugai H (1997) Jpn J Appl Phys 36:337
11. Nagatsu M, Xu G, Ghanashev I, Kanoh M, Sugai H (1997) Plasma Sources Sci Technol 6:427
12. Ghanashev I, Nagatsu M, Xu G, Sugai H (1997) Jpn J Appl Phys 36:4704
13. Sugai H, Ahn TH, Ghanashev I, Goto M, Nagatsu M, Nakamura K, Suzuki K, Toyoda H (1997) Plasma Phys Controlled Fusion 39:A445
14. Ishijima T, Nojiri Y, Toyoda H, Sugai H (2010) Jpn J Appl Phys 49:086002
15. Ishijima T, Toyoda H, Takanishi Y, Sugai H (2011) Jpn J Appl Phys 50:036002
16. Somiya S, Toyoda H, Hotta Y, Sugai H (2004) Jpn J Appl Phys 43:7696
17. Hotta Y, Toyoda H, Sugai H (2007) Thin Solid Films 515:4983
18. Takanishi Y, Okayasu T, Toyoda H, Sugai H (2008) Thin Solid Films 516:3554

19. Kokura H, Yoneda S, Nakamura K, Mitsuhira N, Nakamura M, Sugai H (1999) Jpn J Appl Phys 38:5256
20. Ishikawa K, Ishijima T, Sasai K, Toyoda H, Sugai H (2008) Trans Mater Res Soc Jpn 33:683
21. Takagi Y, Gunjo Y, Toyoda H, Sugai H (2008) Vacuum 83:501
22. Usami K, Ishijima T, Toyoda H (2012) Thin Solid Films 521:22
23. Boisse-Laporte C, Leroy O, de Poucques L, Agius B, Bretagne J, Hugon MC, Teulé-Gay L, Touzeau M (2004) Surf Coat Technol 179:176
24. Thiery F, Pauleau Y, Ortega L (2004) J Vac Sci Technol A22:30
25. Musil J, Mišina M, Hovorka D (1997) J Vac Sci Technol A15:1999
26. de Poucquesa L, Imberta JC, Vasinab P, Boisse-Laportea C, Teulé-Gay L, Bretagnea J, Touzeaua M (2005) Surf Coat Technol 200:800
27. Sasai K, Hagihara T, Noda T, Suzuki H, Toyoda H (2016) Jpn J Appl Phys 55:086202
28. Hasegawa Y, Nakamura K, Lubomirsky D, Park S, Kobayashi S, Sugai H (2017) Jpn. J Appl Phys 56:046203
29. Hottta M, Hasegawa Y, Nakamura K, Kubomirsky D, Park S, Kobayashi S, Sugai J (2017) Jpn J Appl Phys 56:116002 b
30. Moisan M, Zakrzewski Z, Etemadi R, Rostaing JC (1998) J Appl Phys 83:5691
31. Mizojiri T, Morimoto Y, Kando M (2007) Jpn J Appl Phys 46:3573
32. Takamura S, Kando M, Ohno N (2009) J Plasma Fusion Res 8:910
33. Al-Shamma'a AI, Wylie SR, Lucas J, Pau CF (2001) J Phys D 34:2734
34. Iza F, Hopwood JA, Trans IEEE (2003) Plasma Sci 31:782
35. Hoskinson AR, Gregorio J, Parsons S, Hopwood J (2015) J Appl Phys 117:163301
36. Kim JH, Terashima K (2005) Appl Phys Lett 86:191504
37. Schermer S, Bings NH, Bilgic AM, Stonies R, Voges E, Broekaert JAC (2003) Spectrochim Acta, Part B 58:1585
38. Kono A, Sugiyama T, Goto T, Furuhashi H, Uchida Y (2001) Jpn J Appl Phys 40:L238
39. Kono A, Wang J, Aramaki M (2006) Thin Solid Films 506–507:444
40. Itoh H, Kubota Y, Kashiwagi Y, Takeda K, Ishikawa K, Kondo H, Sekine M, Toyoda H, Hori M (2013) J Phys: Conf Ser 441:012019
41. Suzuki H, Nakano S, Itoh H, Sekine M, Hori M, Toyoda H (2015) Appl Phys Express 8:036001
42. Suzuki H, Nakano S, Itoh H, Sekine M, Hori M, Toyoda H (2016) Jpn J Appl Phys 55:01AH09
43. Suzuki H, Toyoda H (2017) Jpn J Appl Phys 56:116001

Chapter 9
Plasma-Assisted Combustion in Automobile Engines Using Semiconductor-Oscillated Microwave Discharge Igniters

Yuji Ikeda

Abstract Plasma-assisted ignition and combustion, and microwave-enhancement of exhaust gas catalytic conversion in automobiles are presented in this chapter. Both processes within and outside of the engine are implemented using a 2.45 GHz semiconductor-oscillated microwave (MW) source or device. The latent potential of fast catalyst light-off was evaluated by applying MW heat to carbon microcoil (CMC) located at the core of a converter. CMC material has high heat absorption efficiency, and is capable of localised and selective heating. The absorption efficiency was approximately 60% with a CMC mass fraction of 10 wt%. Plasma-assisted combustion was implemented by a microwave discharge igniter (MDI) developed by Imagineering, Inc., Japan. The MDI improves combustion performance and enhances lean burn limit due to its ability to generate non-equilibrium, non-thermal plasmas. The oscillator for the MDI has an auto-adjusting capability that ensures the attainment of new resonant frequencies despite the negative effects of material erosion. This allowed continued production of radical species for improving combustion performance. Test runs performed on a commercially available automobile engine demonstrated that the MDI outperforms conventional spark ignition systems.

9.1 Introduction

Currently, about 90% of the world's supply of energy comes from the burning of hydrocarbon/fossil fuels. These are utilised for both air and ground transportation, electric power generation, industrial heating and other industry-related applications [1, 2]. Although hydrocarbon fuels have high energy densities, issues with efficiency and exhaust gas emissions have placed stringent demands on their usage. Regulatory framework to curb the impact of greenhouse gases on climate change, has made it imperative to identify, study and implement new technologies so as to optimize the combustion of hydrocarbon fuels [3–6]. Research studies have shown that automobile exhaust emissions contribute approximately 60–90% of air pollution [6].

Y. Ikeda (✉)
i-Lab., Inc., (formerly Imagineering, Inc.), 7-4-4 Minatojima-Minami, Chuo, Kobe, Japan
e-mail: yuji@i-lab.network

S. Horikoshi and N. Serpone (eds.), *RF Power Semiconductor Generator Application in Heating and Energy Utilization*, https://doi.org/10.1007/978-981-15-3548-2_9

The main composition of the emissions at typical engine operating conditions are carbon monoxide (CO), unburned or partially burned hydrocarbons (HC), nitrogen oxides (NOx) and toxic solid matter known as particulate matter. CO and HC are primarily due to incomplete combustion, while NOx are generated as a result of high combustion temperatures [7–9]. Exhaust gases also contain trace amounts of other harmful substances such as sulphur oxides (SO_x) and phosphorus oxides (PO_x), typically at concentrations below 100 ppm, and thus may not directly pose a toxic risk to the environment. Nevertheless, they act as catalytic impediments (catalytic poisons) that degrade the effectiveness of catalysis when they are adsorbed onto the surface of the catalyst.

Plasma-assisted ignition and combustion are one of the primary measures (within the engine) that generate suitable radicals to enhance combustion chemistry [10–12]. A secondary measure (outside of the engine) is the application of microwave (MW) heating to enhance the catalytic conversion of exhaust gases from toxic substances to less toxic substances such as carbon dioxide (CO_2), water (H_2O) and nitrogen (N_2). In this chapter, we shall discuss first the application of MW heating using a carbon microcoil (CMC) and its effect on the heating system of catalytic converters. Then, we shall describe plasma-assisted combustion implemented by semiconductor-oscillated MW. In particular, we shall examine the extension of the lean limit by a miniaturised device developed by Imagineering, Inc. of Japan termed Microwave Discharge Igniter (MDI). The MDI was applied to a commercially available automobile engine. In both applications, the MW was oscillated by a semiconductor device at a frequency of 2.45 GHz.

9.2 Heating System of Catalytic Converters Using Microwaves [13]

9.2.1 Background

The global energy mix has been projected to be the most diverse by the year 2040 with fossil and non-fossil fuels contributing approximately 25% each. During the said year, renewables are projected to be the fastest-growing source of fuel, contributing about 14% of primary energy. However, the globally projected liquid fuels demand growth for the transport industry is expected to be 1 million barrels per day (Mb/d) for the period 2015–2020, and is expected to decline to approximately one-tenth Mb/d for the period 2030–2035 [14]. In addition, U.S. crude oil and natural gas plant liquids production is projected to grow continuously through 2050 owing to their development of tight oil resources. This would be accompanied by a growth in petroleum product consumption targeted to go beyond 20 Mb/d by 2050 [15]. These projections demonstrate that there will still be demands for fossil-fuelled automobiles especially in developing countries. Thus, catalytic converters with improved functionality to meet the demands of environmental pollution will be a necessity.

Catalytic converters are devices placed at the tailpipe of engine exhaust systems to minimise the emission of toxic substances or to purify exhaust emissions via catalysed redox reactions. They are basically made up of the core, the wash-coat and the catalyst. The core is mostly a ceramic monolith structured like a honeycomb that provides a large surface area (ca. >100 m^2/g) for the processes. Metallic cores are available as well. The wash-coat acts as a support that ensures an effective transfer or scattering of the exhaust substances over the core surface. The wash-coat is usually made of oxides of aluminium, silicon, cerium and titanium or else a mixture of these oxides. The catalyst is usually a combination of platinum-group metals such as rhodium, palladium and platinum for the redox reactions. The most common type of catalytic converters in use are the three-way converters, which are capable of both oxidation and reduction reactions. The main occurring reactions are the oxidation of CO (reaction 9.1) and HC (reaction 9.2), as well as the reduction of NO_x (reaction 9.3):

$$\mathrm{CO} + \frac{1}{2}\mathrm{O_2} \rightarrow \mathrm{CO_2}\ [\text{CO oxidation reaction}] \tag{9.1}$$

$$\mathrm{C}_m\mathrm{H}_n + \left(m + \frac{n}{4}\right)O_2 \rightarrow m\mathrm{CO_2} + \frac{n}{2}(\mathrm{H_2O})\ [\text{HC oxidation reaction}] \tag{9.2}$$

$$\mathrm{NO}_x \rightarrow \frac{1}{2}\mathrm{N_2} + \frac{x}{2}\mathrm{O_2}\ [\mathrm{NO}_x\text{reduction reaction}] \tag{9.3}$$

There have been numerous activities and research efforts to help reduce the toxicity of exhaust emissions. These include, but are not limited to, the development of hybrid-electric vehicles, the use of alternative fuels such as bioethanol and biodiesels, and other methods to warm up the catalytic converter during cold starts. However, these have not significantly mitigated the problems with exhaust emissions. For instance, the frequent switch between electric motor and gasoline engine drives of hybrid automobiles on steep and sloping roads produces higher concentration of CO and HC compared with conventional automobiles. This is because of the cold start of the gasoline engine from a standby state [16]. Moreover, bioethanol engines emit acetaldehyde (CH_3CHO) and higher concentrations of CO, HC and NO_x when compared with conventional engines. Concentrations of NO_x emitted by biodiesel engines are comparatively higher than that of conventional engines [17].

Purification efficiency is high when the catalytic converter operates at relatively high temperatures of about 600 °C and beyond. The nominal time for conventional catalyst light-off is approximately 18–20 s, during which time the front of the converter achieves a temperature of about 300 °C, and the HC emissions reduce from about 2,000 to 1,000 ppm [18]. Consequently, exhaust emissions during the initial start of engines are denser. Some already implemented solutions to ensure the early warm-up of the catalyst include ignition retardation and stratified combustion, a catalytic integrated exhaust manifold, the usage of a pre-catalytic converter, as well as heating directly the catalyst.

In the mid-1990s, a group of German automobile manufacturers developed an electrically heated catalytic converter (EHC) system for the purpose of significantly reducing emissions during the cold start phase of the engine. The EHC system was made up of the following components: a heated catalyst, an electric power supply, an electronic control and a secondary air system [19]. The EHC system developers admitted that the power supply was an area of concern because of its effect on engine management. The EHC system implementation was direct electric heating and thus a dedicated high voltage battery (EHC battery) was required in addition to the vehicle's battery. This certainly added extra mass to the automobile as well as placing constraints on space management. Thus, innovative ways are required to induce early catalyst light-off, in order to significantly reduce the toxicity of automobile exhaust emissions.

We have proposed a system that utilizes a 2.45 GHz MW oscillated by a semiconductor device to heat the carbon microcoil (CMC) [20], which has the advantage of high heating efficiency, as well as the capability of localised and selective heating as the CMC would be located on the front part of the core of catalytic converters. The system is also capable of attaining fast catalyst light-off with temperatures reaching over 250 °C within a few seconds. The proposed system is displayed in Fig. 9.1. The CMC is an amorphous carbon fibre coil with micrometre order pitch and has an excellent electromagnetic absorption property.

The objective for the study of the above-proposed system was to evaluate the latent potential of fast catalyst light-off using a CMC heated by semiconductor-oscillated MW. Semiconductor devices have unique advantages when compared with conventional magnetron devices. Varying degrees of oscillatory patterns are achievable with a semiconductor device, with spatial and temporal control (variable timescales), power levels, and a variable number of pulses per burst. Table 9.1 summarizes the features of both magnetron and semiconductor oscillators. The comparison between oscillatory patterns is illustrated in Fig. 9.2. The experiments were conducted in a constant volume combustion chamber.

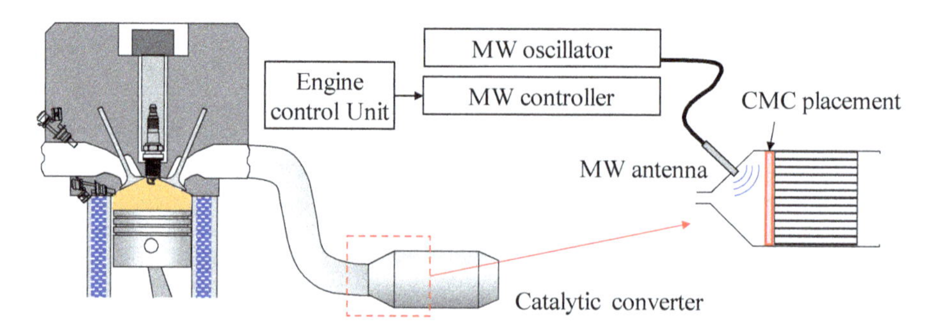

Fig. 9.1 Proposed system for fast catalyst light-off using MW heating with CMC. By Imagineering, Inc., Japan

Table 9.1 Comparison between Magnetron and Semiconductor devices

Parameter	Magnetron	Semiconductor device (Courtesy Ampleon)
Dimensions/mm	80 × 90 × 130	19 × 34 × 4
Efficiency (power/W)	50% (600)	>50% (300)
Lifetime/h	3,000	100,000
control	Non-applicable	Pulse, Phase, frequency, power and feedback control
Response time	300 ms	<1 ns
Noise level/dB	35	<25

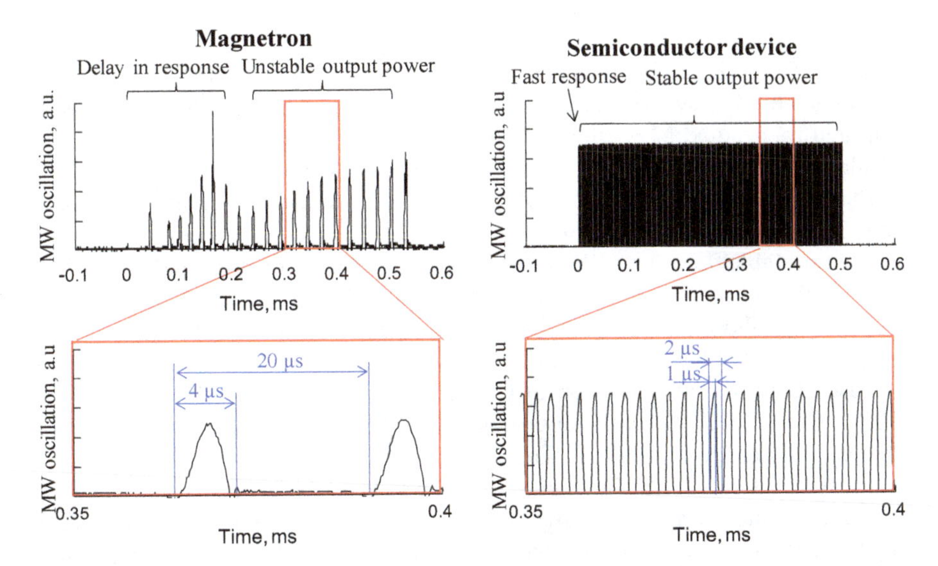

Fig. 9.2 Comparison between the oscillatory patterns of magnetron and semiconductor devices. By Imagineering, Inc., Japan

9.2.2 Experimental Setup and Results

The experimental setup for the measurement of microwave absorption efficiency and temperature are as reported in Fig. 9.3. The test sample, which was composed of varying mixture ratios of a slurry (petalite) and CMC, was placed in the constant volume chamber with a pair of MW antennas connected to the network analyser. The average total mass for each test sample was 26 g. The analyser measured the transmitted and

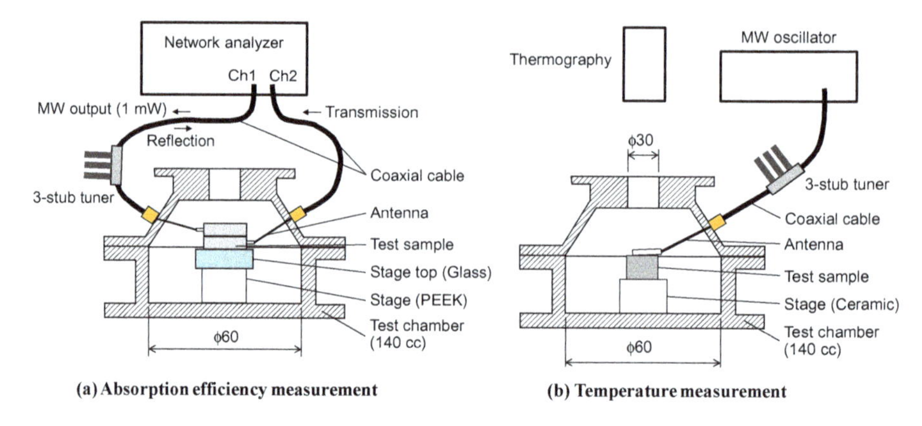

Fig. 9.3 Experimental setup: **a** absorption efficiency measurement, **b** temperature measurement. Reproduced from Ref. [13]. Copyright 2016 by Society of Automotive Engineers of Japan, Inc. (JSAE)

reflected MW, and the stub tuner was used to ensure that effective impedance matching was achieved. The absorbed MW power was estimated as the difference between measured transmitted MW with and without CMC. The absorption efficiency was calculated by comparing the difference to the input MW power.

The surface temperatures of a 100% CMC sample were measured using thermography. The MW oscillation conditions used for the temperature measurements are reported in Table 9.2. The MW was oscillated at a peak power (P_{MW}) of 1 kW, a pulse width (t_W) of 0.5 μs and a pulse period (t_p) of 1 μs.

Figure 9.4 shows the results of absorption efficiency as a function of CMC in weight percent (wt%) in a petalite/CMC mixture. Clearly, the absorption efficiency increased with the mass fraction of CMC. Approximately 10 wt% of CMC was sufficient to achieve effective absorption, since beyond this value the absorption efficiency increment is minimal. Thus, effective heating could be achieved with a relatively small mass of CMC. Figure 9.5 shows the temporal variations of temperature and regions with temperatures beyond 300 °C at a given MW average power. The temperature increases significantly to about 2 s, beyond which the temperature rise is almost constant. It is also observed that at approximately 1 s, an input power of 40 W achieves temperature beyond 300 °C. Thus, the temperature and region of temperature above 300 °C, increase with the MW power.

Table 9.2 MW oscillation conditions for temperature measurements[a]

Average power P_{av}, W	20	25	30	40
MW pulse duration t_{MW}, μs	400	500	600	800

[a]Reproduced from Ref. [13]. Copyright 2016 by Society of Automotive Engineers of Japan, Inc. (JSAE)

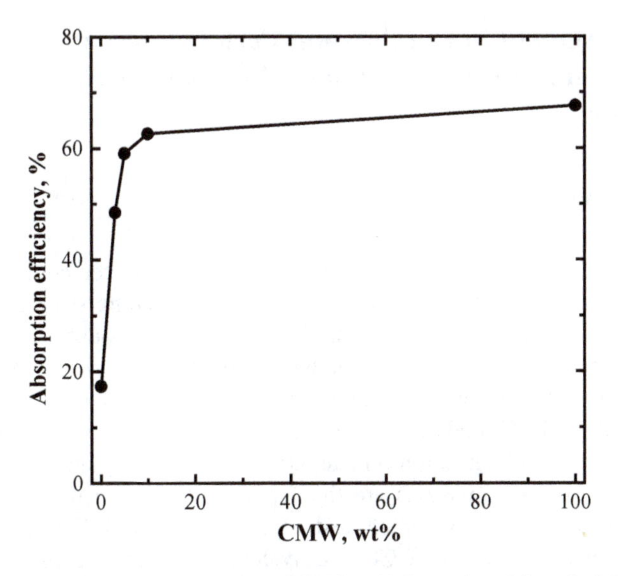

Fig. 9.4 Absorption efficiency as a function of CMC wt%. Reproduced from Ref. [13]. Copyright 2016 by Society of Automotive Engineers of Japan, Inc. (JSAE)

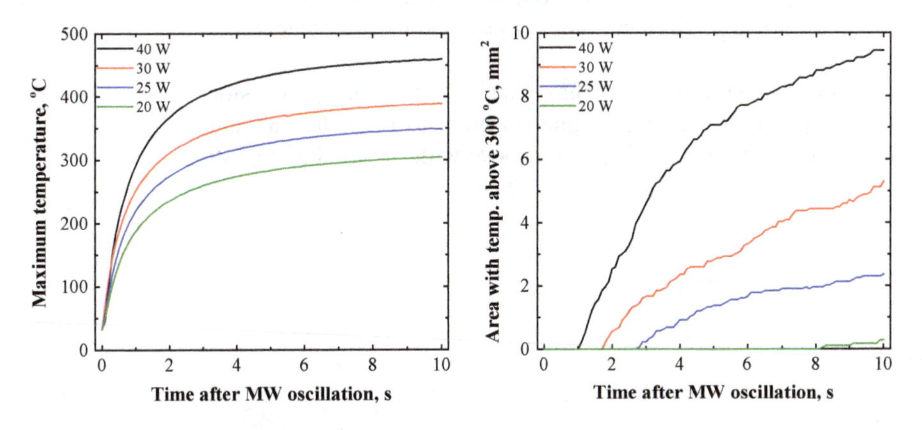

Fig. 9.5 Temporal variations of temperature and region heated beyond 300 °C. Reproduced from ref. [13]. Copyright 2016 by Society of Automotive Engineers of Japan, Inc. (JSAE)

9.2.3 Summary

In this section we evaluated the latent potential of fast catalyst light-off by MW heating applied to CMC using a semiconductor oscillator. The MW absorption efficiency increased significantly to approximately 60% with a CMC mass fraction of 10 wt%. The maximum temperature and region heated beyond 300 °C increased with the MW input power.

9.3 Improvement in Combustion Performance and Lean Burn Limit by a Multi-point Microwave Discharge Igniter

9.3.1 *Background*

As already established in the introduction section of this chapter, countries and international partners at the global level are continuously placing stringent restrictions on passenger and light commercial vehicles almost every consecutive model year. This is evidenced either by the greenhouse gas (GHG) emissions standards of the European Union's (EU) new European driving cycle (NEDC), or the fuel economy-based standards of the United States combined cycle corporate average fuel economy (CAFE) regulations. Other common standards for regulating automobile fuel consumptions/exhaust gas emissions are the Japanese JC08 and the United Nations' world-harmonized light-duty vehicle test procedure (WLTP). For instance, the EU has a set target of 95 grams of CO_2 equivalent per kilometre (g CO_2/km) to be achieved by 2021, while Japan is expected to achieve 82 g CO_2/km by 2020 for passenger vehicles, as well as 133 g CO_2/km for light commercial vehicles. Similarly, the U.S.A. is expected to attain a 45% reduction in GHG emissions from 2010 to 2025. Historical performance and projected targets of other countries are reported in Fig. 9.6 [21].

Numerous research efforts have focused on reducing emissions and maximising efficiency with downsized engines in order to meet the regulatory standards [22]. This is however quite difficult to achieve with conventional ignition systems which

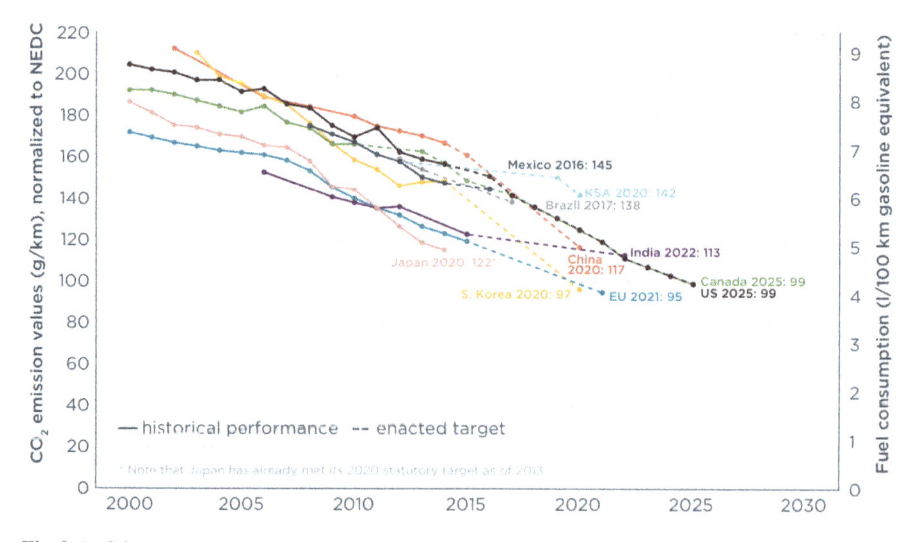

Fig. 9.6 CO_2 emissions performance and present standards normalised to NEDC for passenger cars. Reproduced from Ref. [21]. Copyright 2017 by The International Council on Clean Transportation

operate unsteadily at lean burn and high tumble flow/exhaust gas recirculation (EGR) conditions, due to limited energy supply and durability concerns [23]. Furthermore, successful ignition at maximum EGR conditions requires higher threshold energy values in order to minimise misfiring and flame kernel growth retardation [24, 25]. These barriers have been demonstrably mitigated by the implementation of plasma-assisted ignition (PAI) and plasma-assisted combustion (PAC), which are capable of reaching ultra-lean burn limits while simultaneously reducing exhaust gas emissions [26–33]. Thus, PAI and PAC are viable alternatives to standard spark plug systems.

Imagineering, Inc. of Japan (IMG) has developed several types of Microwave Discharge Igniters (MDIs) capable of breakdown at high ambient pressure for the implementation of both PAI and PAC. The MDI works by the resonation of MW within a quarter coaxial cavity [34–36]. Table 9.3 shows the various models of ignition devices developed and tested by IMG. These were initially built-off the structure of conventional spark plugs. The MDIs are simple, compact in size, and resonated by a 2.45 GHz semiconductor MW oscillator. Various oscillatory patterns are applicable by varying the pulse widths, pulse periods, pulse delay, duty cycle, power per pulse, the number of pulses per burst, and with tens of nanosecond timescale resolutions. Preferential Radical Production (PRP), which is the generation of active radicals, such as O· and ·OH, is implementable by the flexibility in the oscillatory patterns of the MDIs. PRP enhances the low-temperature plasma chemistry that sustains the MW discharge for efficient combustion [37–39]. Thus, the MDI is capable of generating non-equilibrium, non-thermal plasmas which accelerates the growth of flame kernel and reduces misfiring at ultra-lean burn conditions [40–43].

Studies have shown that the single-point MDI enhances combustion compared with standard spark plug ignition systems. Comparatively, the multi-point MDI improves ignition volume, enhances flame development and growth, thereby extending lean burn limits and dilution rates [35, 44–47]. This was demonstrated by the application of an M12, 3-point MDI (each port has a ϕ 4.5 mm plug) to a multi-cylinder commercially available engine. Ignition performance was evaluated and compared to a conventional spark plug ignition system at an engine speed of 1460 rpm and a torque of 20 Nm [48].

9.3.2 Microwave Oscillation and Control

Figure 9.7 shows oscillatory patterns with equal pulse widths (t_W), and equal pulse periods (t_p) labelled as pattern 1, and patterns which have different timescales for set 1 (t_{W1}, t_{p1}) and set 2 (t_{W2}, t_{p2}) pulses labelled as pattern 2 [48]. Oscillation pattern 2 is the most applied pulse train for our MW-enhanced PAC [36, 44]. The MW oscillatory pattern permits the spatial and temporal control of both the expansion and the lifetime of the MW-sustained plasma, which in effect influences the production of active radical species for combustion. The timescales for these oscillation patterns are available from microsecond to nanosecond levels per pulse at MW output power

Table 9.3 Models of MDIs developed and tested by Imagineering, Inc., Japan

Description	Spark ignition (SI)	Spark with MW enhancement (SI + MW)	MDI sustained by MW (MW only)	Multi-point MDI (MW only)		Flat panel igniter	Dual function injector
Structural and discharge image							
Schematic	Ignition coil Spark plug	Ignition coil MW amp. Mixer unit Spark plug (non-resistor/ Straw plug)	MW amp. MDI	MW amp. Co-axial cable	MW amp. M12 plug 4 MDIs in one	MW amp.	MW amp. Co-axial cable Injector Fire-Ring
Development stag	Commercially available	Developed and tested	Developed and tested	Developed and tested	Development phase	Developed and tested	Conceptual phase
Applicable engines	Gasoline and diesel	Gasoline	Gasoline and diesel	Gasoline and diesel	Gasoline and diesel	Gasoline and diesel	Gasoline and diesel

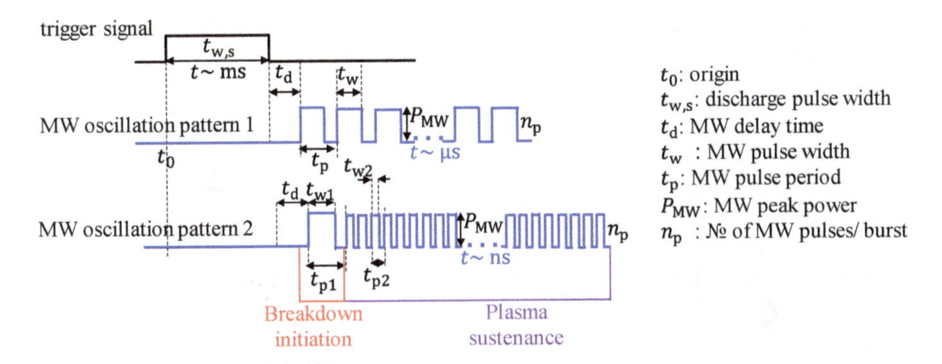

Fig. 9.7 Pictorial representation of MW oscillation pattern and its trigger signal. Reproduced from Ref. [48]. Copyright 2018 by 4th International IAV Conference

(P_{MW}) of tens of Watt to kiloWatt levels per pulse, thus achieving several hundreds of milliJoules per pulse. The available number of pulses per burst (n_p) is up to 2400.

The equivalent electric circuit for a single cavity as well as pictures of 1-point and 3-point MDI with MW-enhanced plasma during discharge are shown respectively in Figs. 9.8a, b [36]. The 3-point MDI could be operated as a single to a 3-port discharge device, since each of the ports is separately controlled. In Table 9.4, we present oscillatory patterns previously studied and how they affect the generation of O· radicals are shown in Fig. 9.9 [48].

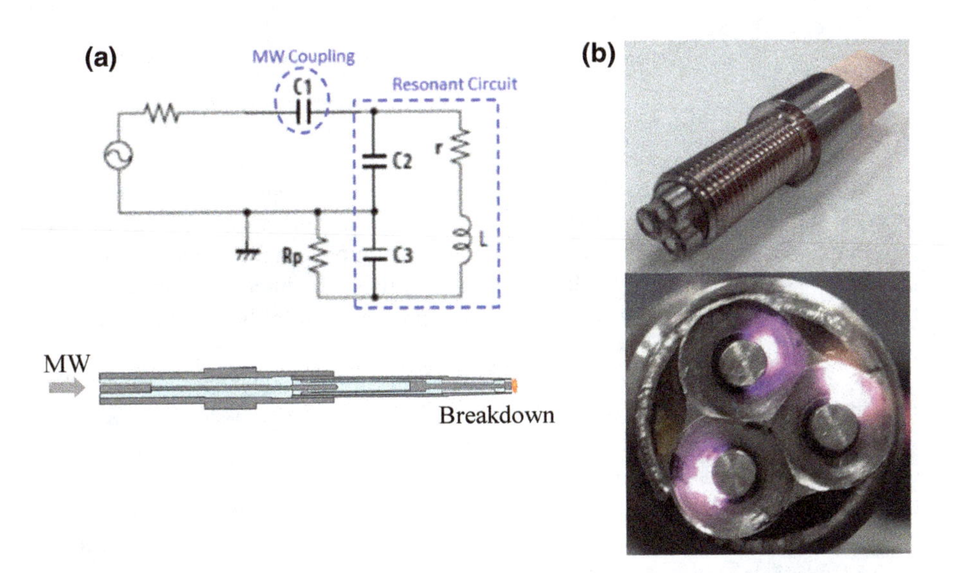

Fig. 9.8 **a** Equivalent electric circuit and image of a single cavity MDI, **b** pictures of the M12, 3-point MDI with MW-enhanced discharge. Reproduced from ref. [36]. Copyright 2018 by 4th International IAV Conference

Table 9.4 MW oscillation used for O· radical generation shown in Fig. 9.9 (E_{MW} is the MW input energy per total period).[a]

Pattern №	MW Oscillation Pattern			
	t_{p2}/ms	t_{w2}/ms	n_{p2}	E_{MW}/mJ
1	2	0.1	1400	232
2	2	0.3	700	344
3	2	0.2	1400	456

[a]Reproduced from Ref. [48]. Copyright 2018 by 4th International IAV Conference

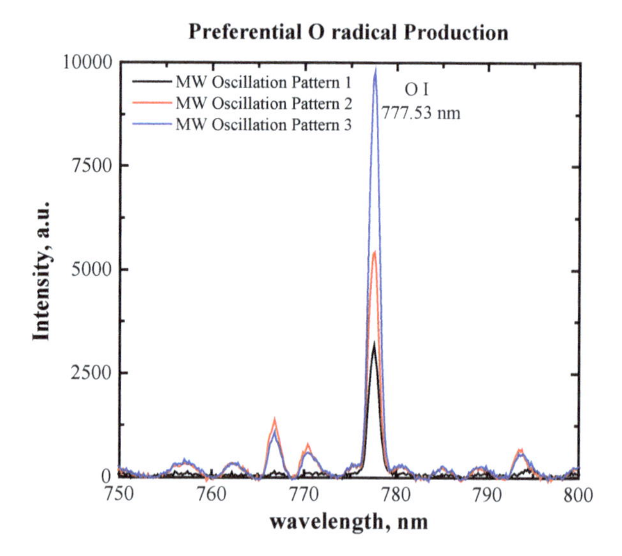

Fig. 9.9 The effect of MW oscillation pattern on O· radical production. Reproduced from Ref. [48]. Copyright 2018 by 4th International IAV Conference

The O· radical is a vital intermediate species during flame development and growth, and thus knowing the oscillatory patterns that influence its production is important. We observe in Fig. 9.9 that the oscillation pattern 3 (blue spectrum) has the highest spectral intensity, which is very intuitive. The intensity corresponds to the input MW energy, which was high and is proportional to the number of pulses per burst.

The effect of the oscillatory pattern per pulse on plasma characteristics are displayed in Fig. 9.10 [44]. Longer t_W are required to induce breakdown, as a result of which induces the erosion of the electrode and the antenna material. This is evidenced by the atomic lines of pulse 1 in Fig. 9.10. In contrast, shorter t_W are enough to sustain the induced plasma, which has the advantage of minimising erosion while generating molecular spectra and radicals. This can also be seen in the subsequent pulses (pulses 3–5) of Fig. 9.10.

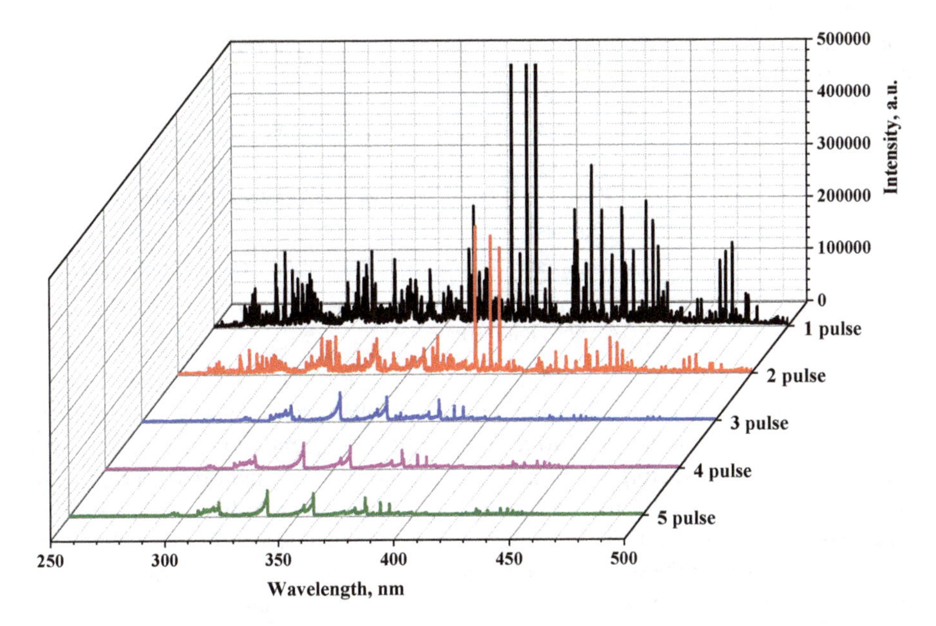

Fig. 9.10 The effect of MW oscillation pattern per pulse on plasma characteristics. Reproduced from Ref. [44]. Copyright 2017 by SAE Technical Paper

9.3.3 MDI Performance Test with Constant Volume Combustion Chamber

Stress and endurance tests of a single-point MDI were performed in a constant volume combustion chamber (CVCC) [34–36]. The CVCC used for the study is a pent-roof type with a capacity of 185 mL and is supplied with a propane-air mixture from a pre-mixing tank. The temperature of both the chamber and its contents were set at 298 K during the test-runs. The MDI was installed in the centre of the pent-roof and the chamber was equipped with a piezo-electric transducer (Kistler 6052 C) for pressure measurements. The MDI device was test-ran first, continuously for 10 h at wide-open-throttle load conditions and then continuously for 100 h at the Indicated Mean Effective Pressure (IMEP) of 550 kPa part load condition. These test-runs were carried out at engine speeds of 3000 rpm. For these test conditions, the MDI was robust, flexible and also outperformed the standard spark ignition system.

The stress tests were conducted in a non-reactive medium at a pressure of 0.5 MPa. The MDI was made to discharge continuously for 124 h over 20 million discharge counts replicating an equivalence of a standard lifetime of 20,000 km highway driving. The experimental conditions at which the stress tests were performed are summarised in Table 9.5. These tests led to the erosion of the electrode material as well deposition of substances on the outer parts of the MDI. The erosion and deposition have the potential to restructure the MDI's geometry, thereby influencing its resonance performance. Pictures of the MDI before, during, and after the stress tests are

Table 9.5 Experiment conditions of the stress test in the CVCC[a]

Parameter	Condition
№ of discharges	Over 20 million times
Discharge frequency	50 Hz (equivalent to an ignition interval of 6000 rpm)
Gas	N_2 (non-reactive, no oxidation)
Pressure	0.5 MPa
Flow rate	<1000 mL/min (for gas exchange)

[a]Reproduced from ref. [36]. Copyright 2018 by Proceedings of the Combustion Institute

illustrated in Fig. 9.11 [36]. The semiconductor MW oscillator was able to monitor, self-correct and re-adjusted to a new resonant frequency to ensure successful ignition, thereby preventing misfiring. The self-adjusted resonant frequencies were measured by an RF vector network analyser and are shown in Fig. 9.12 [48]. This simply means that the MDI has the capacity to withstand tens of millions of discharge counts, during which the semiconductor oscillator induces waves at frequencies that ensure efficient coupling with the plasma allowing high combustion efficiency.

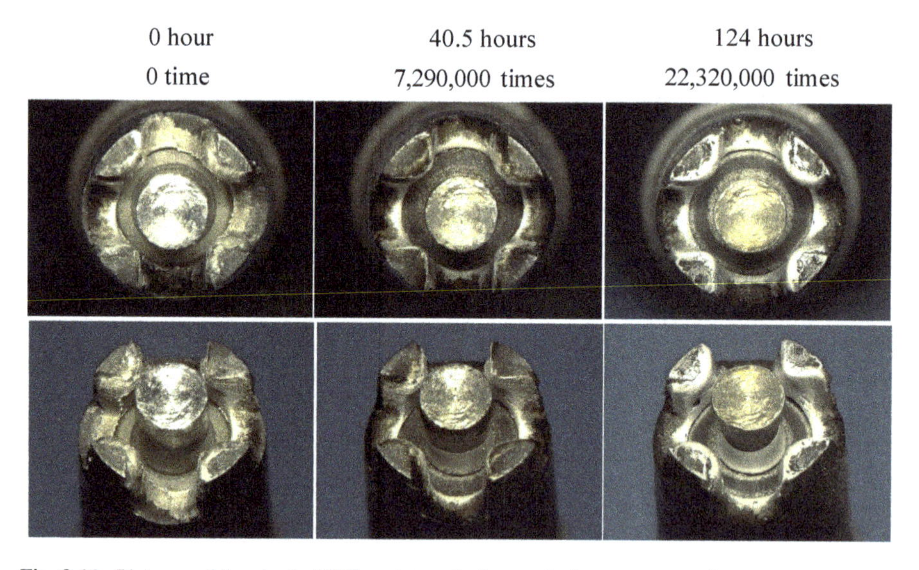

Fig. 9.11 Pictures of the single MDI prototype before and after stress tests in a constant volume chamber (top: front view, bottom: tilted angle view). Reproduced from Ref. [36]. Copyright 2018 by Proceedings of the Combustion Institute

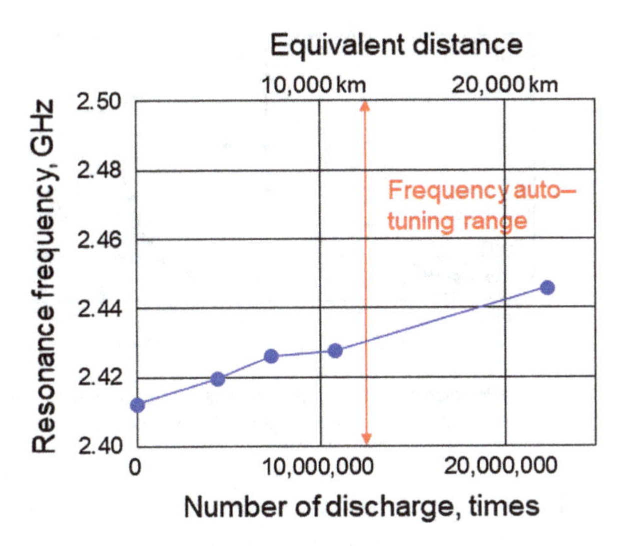

Fig. 9.12 Auto frequency adjustment during the stress test in the CVCC. Reproduced from Ref. [48]. Copyright 2018 by 4th International IAV Conference

9.3.4 MDI Performance Test with Multi-cylinder Engine

The engine used for the test is a commercially available, naturally aspirated three-cylinder Daihatsu KF-VE5, with a port-fuel-injection system for the intake. The specifications of the engine are presented in Table 9.6. The engine was not optimally designed for higher lean limits and dilution rates. Hence, all tests performed were conducted at an engine speed of 1460 rpm and a torque of 20 Nm. Data recording was done by a standard data logging system provided by Daihatsu. The test [15] showed that the multi-point MDI outperformed the standard spark ignition system in all three cylinders of the production engine. Furthermore, the 2-point experimented MDI exhibited improved performance when compared with the single-point experimented MDI, attaining the air-fuel ratio (AFR) of 31. Although the engine was strictly not

Table 9.6 Specifications and operating conditions of the multi-cylinder engine

Parameter	Specification
№ of cylinders	3
Engine model	KF-VE5
Displacement	658 mL
Compression ratio	12.2
Bore × stroke, mm	63 × 70.4
Fuel injection	PFI
Engine speed	1460 rpm
Torque	20 Nm
EGR rate	0
Excess air ratio	>2.1

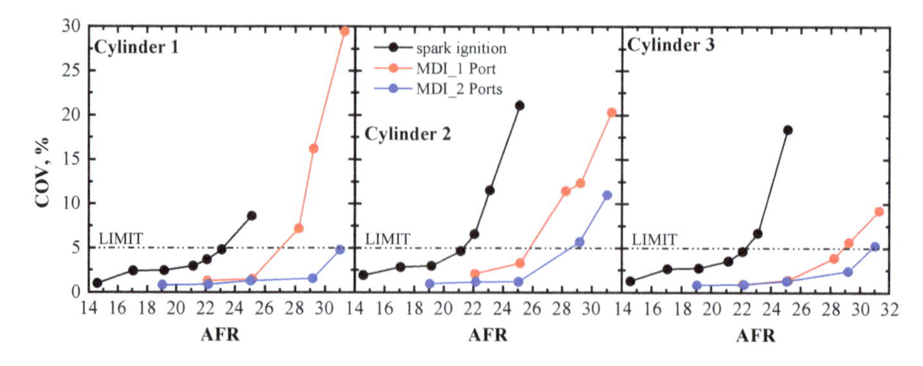

Fig. 9.13 Performance comparison of spark plug and 1/2-point MDI in multi-cylinder engine. Reproduced from Ref. [36]. Copyright 2018 by Proceedings of the Combustion Institute

optimally designed for higher lean limits at the studied low speed and low torque conditions, multi-point MDI showed a better performance. We observe from Fig. 9.13 [36] that the coefficient of variation (COV) for cylinder 2 was the largest (worst) in all cases and conditions of the study. This was caused by a cylinder difference of combustion owing to the mounting platform constraint imposed on the intake manifolds.

9.3.5 Load Performance Test of Lean Burn

The load performance, or lean limit expansion, test results are shown for cylinder 1 of the multi-cylinder engine. The variations of the IMEP with crank angle (CA) during the flame development time, that is, between discharge initiation and 10% of cumulative net heat release are given in Fig. 9.14 [48]. The as-displayed IMEP variations are for 300-cycle test-runs each for selected AFRs up to the limiting conditions for the spark 1- and 2-point MDI ignition systems, respectively. We observe that under leaner conditions, a long initial combustion period generates low IMEP cycle. The standard spark ignition system is unstable at AFR of 25, while the 1-point MDI becomes unstable beyond AFR of 25. The 2-point MDI performed comparatively better up to an AFR of 31. This emphasizes the effectiveness of the multi-point MDI for enhancing the efficiency of combustion with MW. It should be noted that the compactness of the multi-point MDI minimises flexibility concerns. This makes it applicable to various designs and configurations of combustion chambers with different engine heads.

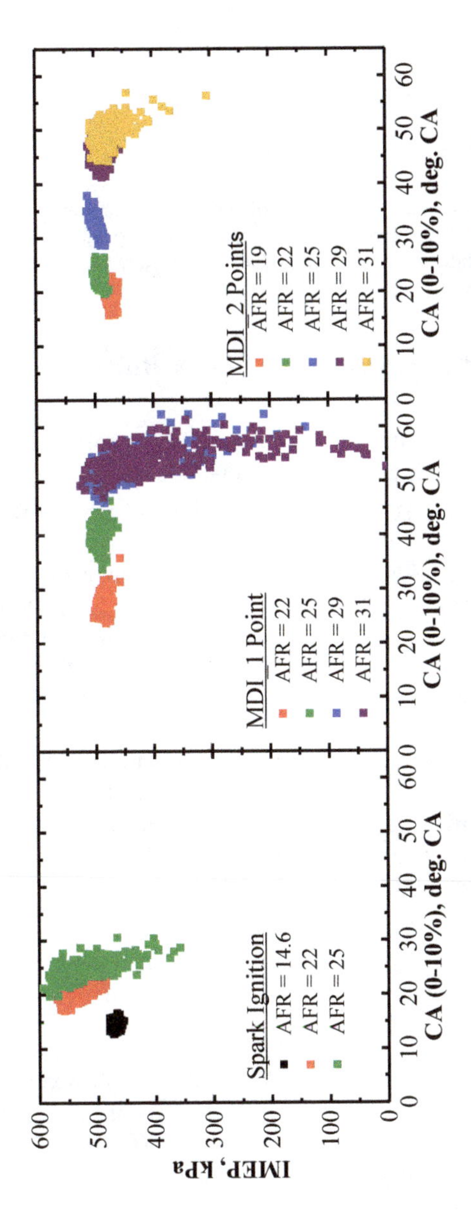

Fig. 9.14 IMEP variation as a function of CA during flame development time for the ignition systems studied (Load Performance Test results for cylinder 1). Reproduced from Ref. [48]. Copyright 2018 by 4th International IAV Conference

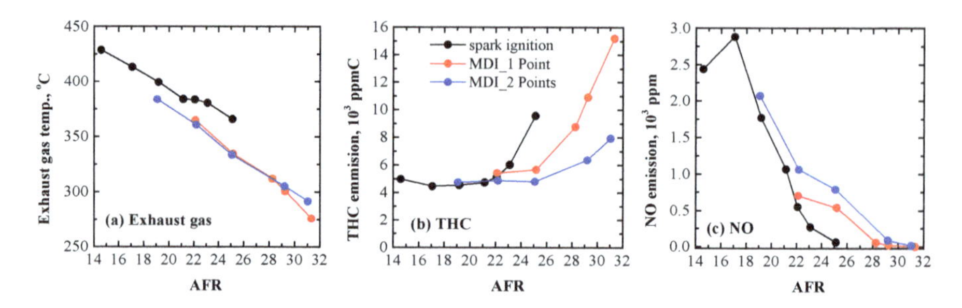

Fig. 9.15 Exhaust gas characteristics as a function of AFR for the ignition systems studied (Emission Performance Test results for cylinder 1): **a** Exhaust gas temperature, **b** Total hydrocarbon emission, and **c** Nitrogen oxides emission. Reproduced from Ref. [48]. Copyright 2018 by 4th International IAV Conference

9.3.6 Emission Performance Test of Lean Burn

We demonstrate in this section that the MDI and the multi-point MDI exhibit a superior and improved exhaust gas emission characteristic when compared to the standard spark plug ignition systems. Figures 9.15a–c show, respectively, the exhaust gas temperatures (before catalyst), the total hydrocarbon (THC), and the NO emissions for cylinder 1 for all the ignition systems studied [48]. The MDI outperformed the standard spark ignition system, exhausting at relatively low temperatures at comparatively high AFR values. More significant, the multi-point MDI maintained exhaust gas temperatures below 300 °C at an AFR value of 31. The THC emission of the 2-point MDI was the minimum for lean burn conditions. The multi-point MDI emitted single digit ppm of NO_x compounds at the AFR values of 30 and 31, although the NO_x emissions were the highest for the 2-point MDI within AFR values of 20–25. Therefore, the MW-assisted combustion is a robust and an efficient system compared to combustion via standard spark ignition. The basic reason is the capability of the multi-point MDI to generate non-equilibrium plasma with active radical species that enhance the ignition and combustion processes.

9.3.7 Summary

We have demonstrated in this section that PAI and PAC implemented by MDIs have improved performance characteristics than those of standard plug ignition systems. We presented the semiconductor MW oscillator as a versatile device, capable of an auto-adjusting mechanism despite the negative effects of erosion and deposition of metal electrode materials on cavity structure. This allowed new resonant frequencies to be achieved that ensured radical species production, enhanced low-temperature plasma chemistry, and hence improved combustion efficiency.

Considering the test-runs performed on the commercially available multi-cylinder engine (Daihatsu Motors), the multi-point MDI exhibited an improved COV of IMEP performance compared to the standard spark ignition system. The multi-point MDI extended the lean limit expansion to an AFR of 31 (i.e., lambda, λ, of 2.1) showing a comparatively better performance than that of the 1-point MDI and spark ignition. Thus, MW coupling for increased combustion efficiency is demonstrably possible with the multi-point MDI device in practical automobile engines.

Finally, at higher lean burn limits, the multi-point MDI maintained exhaust gas temperatures below 300 °C, and achieved lower THC and NO_x emissions. Thus, the multi-point MDI is one of the potential candidate devices that could assist the attainment of targets continuously imposed by the stringent emission regulatory standards for automobiles.

9.4 A Novel Plasma Igniter

IMG has tested and is currently improving on the design of a miniaturised novel igniter, termed as the Flat-Panel Plasma Igniter (FPI). The FPI is an 8 mm × 8 mm × 0.4 mm ceramic panel with conductive inlay for microwave resonation. The image of the FPI with MW-induced and sustained discharge are shown in Fig. 9.16 [49]. Performance studies of the FPI have been conducted in a CVCC with propane-air mixture where successful ignition was achieved at an equivalence ratio of 0.6. The FPI's size is a characteristic advantage for various geometries of combustion chambers. We are of the view that this device will revolutionize the automobile industry in terms of the objective of achieving reduced emissions with highly efficient combustion engines.

Fig. 9.16 Miniaturised flat-panel plasma igniter with MW-induced and sustained discharge. Reproduced from Ref. [49]. Copyright 2017 by SAE Technical Paper

References

1. Chu S, Majumdar A (2012) Opportunities and challenges for a sustainable energy future. Nature 488:294–303. https://doi.org/10.1038/nature11475
2. Warnatz J, Maas U, Dibble RW (2006) Combustion, physical and chemical fundamentals, modelling and simulation, experiments, pollutant formation, 4th edn. Springer, Berlin, p 1
3. Global fuel economy initiative, *Delivering Climate Action Report* (2018). https://www.globalfueleconomy.org/. Accessed Nov 2018
4. Dec JE (2009) Advanced compression-ignition engines-understanding the in-cylinder processes. Proc Combust Inst 32:2727–2742. https://doi.org/10.1016/j.proci.2008.08.008
5. Christensen M, Hultqvist A, Johansson B (1999) Demonstrating the multi fuel capability of a homogeneous charge compression ignition engine with variable compression ratio. SAE Trans 108:2099–2113. http://www.jstor.org/stable/44743532
6. Chen J, Zu K, Wang M (2018) Experimental study on the plasma purification for diesel engine exhaust gas. IOP Conf Ser Earth Environ Sci 113:012186
7. Jones JL, Sckuck EA, Eldridge RW, Endow N, Cranz FW (1963) Measurement of automobile exhaust gas hydrocarbons. J Air Pollut Control Assoc. 13:73–77. https://doi.org/10.1080/00022470.1963.10468144
8. Baker RA Sr, Doerr RC (1964) Catalytic reduction of nitrogen oxides in automobile exhaust. J Air Pollut Control Assoc 14:404–409. https://doi.org/10.1080/00022470.1964.10468305
9. Trinh HT, Imanishi K, Morikawa T, Hagino H, Takenaka N (2017) Gaseous nitrous acid (HONO) and nitrogen oxides (NO_x) emission from gasoline and diesel vehicles under real-world driving test cycles. J Air Waste Manage Assoc 67:412–420. https://doi.org/10.1080/10962247.2016.1240726
10. Starikovskaia SM (2006) Plasma assisted ignition and combustion. J Phys D Appl Phys 39:R265–R299. https://doi.org/10.1088/0022-3727/39/16/R01
11. Starikovskaia SM (2014) Plasma assisted ignition and combustion: nanosecond discharges and development of kinetic mechanisms. J Phys D Appl Phys 47:353001. https://doi.org/10.1088/0022-3727/47/35/353001
12. Ju Y, Sun W (2015) Plasma assisted combustion: dynamics and chemistry. Progr Energy Combust Sci 48:21–83. https://doi.org/10.1016/j.pecs.2014.12.002
13. Ikeda Y, Nishiyama A, Wachi Y (2016) Heating system of catalytic converter using the semiconductor microwave power supply. In: Society of automotive engineers of Japan, Inc. (JSAE) proceedings, No. 105-16, Oct 2016, pp 189–194 (In Japanese)
14. Edition of BP Energy Outlook, BP Energy Economics, pp 67–80. https://www.bp.com/en/global/corporate/energy-economics.html. Accessed Oct 2018
15. Annual Energy Outlook 2018, with Projections to 2050, U.S. Energy Information Administration https://www.eia.gov/outlooks/aeo/. Accessed Oct 2018
16. Martini G, Bonnel P, Manfredi U, Carriero M, Krasenbrink A, Franken O, Rubino L, Bartoli GB, Bonifacio M (2010) On-road emissions of conventional and hybrid vehicles running on neat or fossil fuel blended alternative fuels. SAE Technical Paper 2010-01-1068. https://doi.org/10.4271/2010-01-1068
17. Kawano D, Ishii H, Goto Y (2008) Effect of biodiesel blending on emission characteristics of modern diesel engine. SAE Technical Paper 2008-01-2384. https://doi.org/10.4271/2008-01-2384
18. Baba N, Osawa K, Sugiura S (1996) Analysis of transient thermal and conversion characteristics of catalytic converters during warm-up. JSAE Trans 27:59–65 (in Japanese)
19. Pfalzgraf B, Otto E, Wirth A, Küper P, Held W, Donnerstag A (1995) The system development of electrically heated catalyst (EHC) for the LEV and EU-III legislation. SAE Technical Paper 951072. https://doi.org/10.4271/951072
20. Motojima S, Chen X (2007) Preparation and characterization of carbon microcoils (CMCs). Chem Soc Jpn 80:449–455. https://doi.org/10.1246/bcsj.80.449

21. Yang Z, Bandivadekar A (2017) Global update: light-duty vehicle, greenhouse gas, and fuel economy standards. In: The international council on clean transportation, pp 1–27. https://www.theicct.org/. Accessed Oct 2018
22. Johnson T, Joshi A (2017) Review of vehicle engine efficiency and emissions. SAE Technical Paper 2017-01-0907. https://doi.org/10.4271/2017-01-0907
23. Hakariya M, Toda T, Sakai M (2017) The new Toyota inline 4-cylinder 2.5L gasoline engine. SAE Technical Paper 2017-01-1021. https://doi.org/10.4271/2017-01-1021
24. Yao M, Zhang Q, Liu H, Zheng Z, Zhang P, Lin Z, Lin T, Shen J (2010) Diesel engine combustion control: medium or heavy EGR? SAE Technical Paper 2010-01-1125. https://doi.org/10.4271/2010-01-1125
25. Tromans, P. S., Furzeland R. M. An analysis of Lewis number and flow effects on the ignition of premixed gases, *Symposium (International) on Combustion* 21 (1988) 1891-1897. https://doi.org/10.1016/S0082-0784(88)80425-9
26. Ikeda Y, Padala S, Makita M, Nishiyama A (2015) Development of innovative microwave plasma ignition system with compact microwave discharge igniter. SAE Technical Paper 2015-24-2434. https://doi.org/10.4271/2015-24-2434
27. Wolk B, DeFilippo A, Chen JY, Dibble R, Nishiyama A, Ikeda Y (2013) Enhancement of flame development by microwave-assisted spark ignition in constant volume combustion chamber. Combust Flame 160:1225–1234. https://doi.org/10.1016/j.combustflame.2013.02.004
28. Ma JX, Alexander DR, Poulain DE (1998) Laser spark ignition and combustion characteristics of methane-air mixtures. Combust Flame 112:492–506. https://doi.org/10.1016/S0010-2180(97)00138-7
29. Shiraishi T, Urushihara T, Gundersen MA (2009) A trial of ignition innovation of gasoline engine by nanosecond pulsed low temperature plasma ignition. J Phys D Appl Phys 42:135208. https://doi.org/10.1088/0022-3727/42/13/135208
30. Kim HH, Takashima K, Katsura S, Mizuno A (2001) Low-temperature NO_x reduction processes using combined systems of pulsed corona discharge and catalysts. J Phys D Appl Phys 34:604–613. https://doi.org/10.1088/0022-3727/34/4/322
31. Khacef A, Cormier JM, Pouvesle JM (2002) NO_x remediation in oxygen-rich exhaust gas using atmospheric pressure non-thermal plasma generated by a pulsed nanosecond dielectric barrier discharge. J Phys D Appl Phys 35:1491–1498. https://doi.org/10.1088/0022-3727/35/13/307
32. Czemichowski A (1994) Gliding arc. applications to engineering and environment control. Pure Appl Chem 66:1301–1310
33. Shiraishi T (2012) Possibility of the new ignition system using the low temperature plasma having dual functions of strengthening ignition for si combustion and promoting and controlling auto ignition of HCCI combustion. In: 1st IAV international conference on advance ignition systems for gasoline engines, Expert Verlag, Berlin, pp 82–94, 12–13 Nov 2012
34. Le M, Padala S, Nishiyama A, Ikeda Y (2017) Control of microwave plasma for ignition enhancement using microwave discharge igniter. SAE Technical Paper 2017-24-0156. https://doi.org/10.4271/2017-24-0156
35. Padala S, Nagaraja S, Ikeda Y, Le M (2017) Extension of dilution limit in propane-air mixtures using microwave discharge igniter. SAE Technical Paper 2017-24-0148. https://doi.org/10.4271/2017-24-0148
36. Le M, Nishiyama A, Serizawa T, Ikeda Y (2018) Applications of a multi-point microwave discharge igniter in a multi-cylinder gasoline engine. In: Proceedings of the combustion institute (In Press, Corrected Proof). Available online 3 July 2018. https://doi.org/10.1016/j.proci.2018.06.033
37. Moisan M, Zakrzewski Z (1991) Plasma sources based on the propagation of electromagnetic surface waves. J Phys D Appl Phys 24:1025–1048. https://doi.org/10.1088/0022-3727/24/7/001
38. Gerstein M, Choudhury PR (1985) Use of silane-methane mixtures for scramjet ignition. J Propul Power 1:399–402
39. Wang F, Liu JB, Sinibaldi J, Brophy C, Kuthi A, Jiang C, Ronney P, Gundersen MA (2005) Transient plasma ignition of quiescent and flowing air/fuel mixtures. IEEE Trans Plasma Sci 33:844–849. https://doi.org/10.1109/TPS.2005.845251

40. Potts H, Hugill J (2000) Studies of high-pressure, partially ionized plasma generated by 2.45 GHz microwaves. Plasma Sour Sci Technol 9:18–24. https://doi.org/10.1088/0963-0252/9/1/304
41. Ogura K, Yamada H, Sato Y, Okamoto Y (1997) Excitation temperature in high-power nitrogen microwave-induced plasma at atmospheric pressure. Appl Spectros 51:1496–1499. https://doi.org/10.1366/0003702971938984
42. Laroussi M, Roth JR (1993) Numerical calculation of the reflection, absorption, and transmission of microwaves by a nonuniform plasma slab. IEEE Trans Plasma Sci 21:366–372. https://doi.org/10.1109/27.234562
43. Ju Y, Guo H, Maruta K, Liu F (1997) On the extinction limit and flammability limit of non-adiabatic stretched methane-air premixed flames. J Fluid Mechan 342:315–334. https://doi.org/10.1017/S0022112097005636
44. Shcherbanev S, De Martino A, Khomenko A, Starikovskaia S, Padala S, Ikeda Y (2017) Emission spectroscopy study of the microwave discharge igniter. SAE Technical Paper 2017-24-0153. https://doi.org/10.4271/2017-24-0153
45. Zheng M, Yu S, Tjong J (2017) High energy multipole distribution spark ignition system. In: 3rd IAV international conference on advance ignition systems for gasoline engines. Springer International Publishing, Switzerland, pp 109–130
46. Morsy MH, Chung SH (2003) Laser-induced multi-point ignition with a single-shot laser using two conical cavities for hydrogen/air mixture. Exp Therm Fluid Sci 27:491–497. https://doi.org/10.1016/2FS0894-1777(02)00252-2
47. Ronney PD (1994) Laser versus conventional ignition of flames. Opt Eng 33:33–44. https://doi.org/10.1117/12.152237
48. Nishiyama A, Ikeda Y, Serizawa T (2018) Lean limit expansion up to lambda 2 by multi-point microwave discharge igniter. In: Ignition systems for gasoline engines. In: 4th international IAV conference. Berlin, Germany (Submitted), 6–7 Dec 2018
49. Padala S, Le M, Nishiyama A, Ikeda Y (2017) Ignition of propane-air mixtures by miniaturized resonating microwave flat-panel plasma igniter. In: SAE Technical Paper 2017-24-0150. https://doi.org/10.4271/2017-24-0150

Part IV
New Application

Chapter 10
Microwave-Assisted Magnetic Recording

Satoshi Okamoto

Abstract Microwave-assisted magnetic recording (MAMR) has attracted much attention as one of the promising next-generation ultra-high density recording technologies of hard disk drives (HDDs). MAMR is not only the technology which simply extends the recording density of HDDs but also has the potential of 3D magnetic recording. In this section, in addition to the explanations of the basic mechanism and some of the demonstrations of MAMR, some efforts for 3D magnetic recording based on MAMR technology will be explained.

10.1 Introduction

At the present time, the value of big data has been widely recognized in various fields. Data science has contributed significantly to solutions of many problems that have been difficult to deal in the past. One of the critical technologies supporting the big data era is the data storage system. The hard disk drive (HDD, see Fig. 10.1), has been a critical device in modern data storage systems primarily because of its extremely high recording density, very low bit cost, long-term data retention, random access, and high-speed data transfer rate. As shown in Fig. 10.2, flash memory has also been increasing in its recording density, and its percentage in the total data storage will rapidly increase. Actually, we have experienced that HDDs were replaced by flash memories in mobile devices, laptop computers, and so on. However, when we focus on the data storage in data center, HDD has kept a major role.

The recording density of HDDs has increased constantly year by year to meet the rapid increase in data traffic. However, the recent increment in recording density of HDDs has plateaued owing to some technological problems. To overcome these problems, many new recording technologies for the next-generation ultra-high-density HDDs have been proposed, such as heat-assisted magnetic recording (HAMR) and bit-patterned magnetic recording (BPMR) [1]. However, it has been very difficult to

S. Okamoto (✉)
Institute of Multidisciplinary Research for Advanced Materials (IMRAM), Tohoku University, Sendai 980-8577, Japan
e-mail: satoshi.okamoto.c1@tohoku.ac.jp

S. Horikoshi and N. Serpone (eds.), *RF Power Semiconductor Generator Application in Heating and Energy Utilization*, https://doi.org/10.1007/978-981-15-3548-2_10

Fig. 10.1 Photograph of a HDD. Copyright 2018 by Satoshi Okamoto

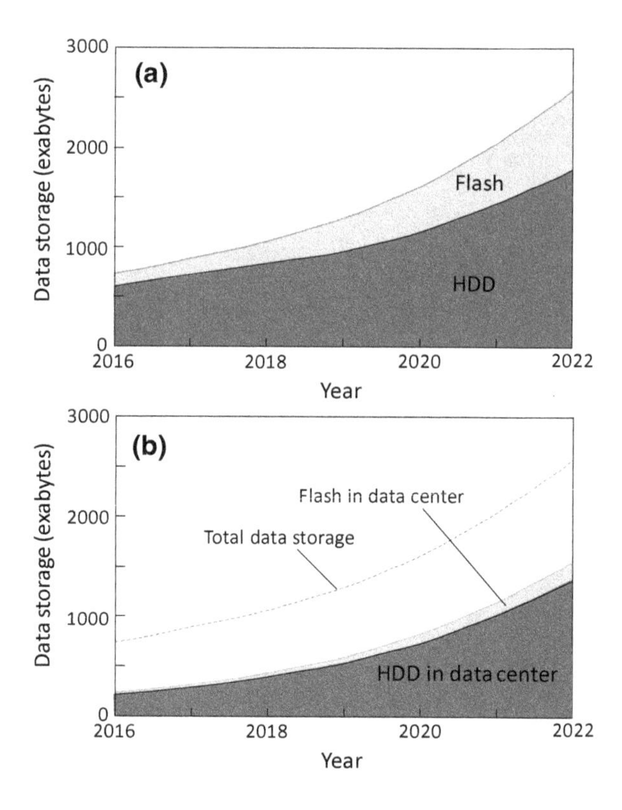

Fig. 10.2 Year trend of data storage. **a** Total data storage and **b** that in data center. Date source is the 2016 press release from Western digital company. Copyright 2018 by Satoshi Okamoto

maintain low bit cost, high reliability, and technological succession in these proposed technologies. Microwave-assisted magnetic recording (MAMR) is a new technology that was proposed in 2008 [2]; it has become the most promising candidate because of its ease of application in conventional recording systems without requiring significant system changes. Figure 10.3 shows the technology road map.

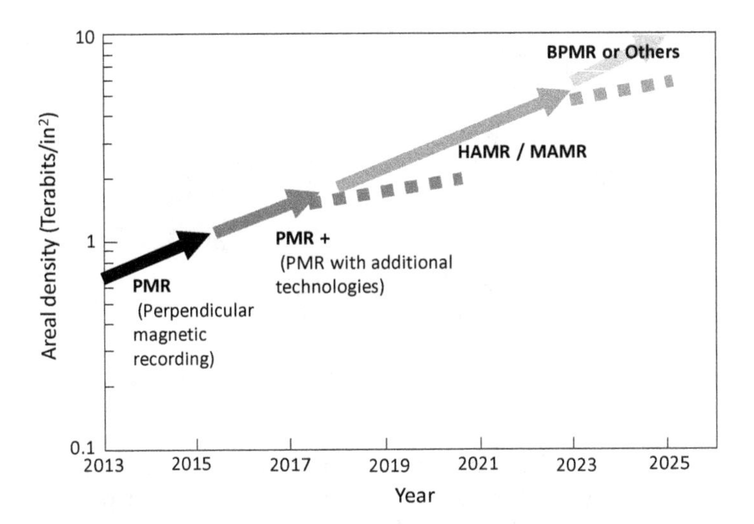

Fig. 10.3 Technology road map of HDDs. Date source is the 2016 report from ASTC/ASRC. Copyright 2018 by Satoshi Okamoto

HDDs contain a very thin magnetic film medium (ca. 10 nm in thickness) that is deposited on a disk substrate. The disk is rotated at several thousand rpm. An HDD head consisting of ***write*** and ***read*** parts is loaded onto the rotating medium with the spacing between the head and the medium surface kept to a few nm as a result of the air–fluid pressure (air bearing). The ***write*** part consists of a write pole of magnetic material with high saturation magnetization, an excitation coil, and a magnetic shield surrounding the write pole. The ***read*** part consists of a highly sensitive magnetic field sensor and a magnetic shield. By applying a current through the excitation coil, a write field is generated from the write pole, after which the magnetization of the medium is switched to form a recorded bit pattern. To increase the recording density further, a medium material with large switching field is indispensable because of high tolerance against thermal agitation. The upper limit of the write field, however, is governed by the saturation magnetization of the write pole material. The latest HDDs are already utilizing the write pole material, which has the highest saturation magnetization among available magnetic materials. It is, therefore, impossible to increase the write field further.

Microwave-assisted magnetic recording is a potential solution that addresses the write field limitation problem, which allows the medium switching field to be significantly decreased by applying a large amplitude microwave field. From a technological point of view, there are two important issues on MAMR, namely a magnetization switching under a microwave field and developing a way to generate a microwave field inside a very small recording head. The former is referred to as microwave-assisted switching (MAS) and has been verified extensively by model experiments [3–6] and micromagnetic computer simulations [7–9]. The proposed solution to the latter problem is to use a spin-torque oscillator (STO) [2, 10–12]. Figure 10.4 shows

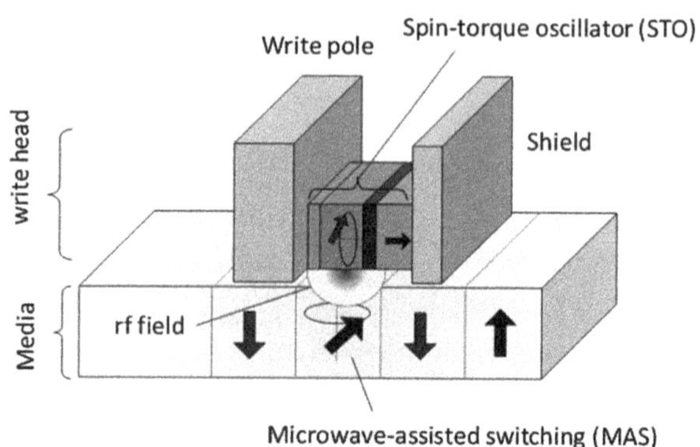

Fig. 10.4 Schematic illustration of write head and medium for MAMR. Copyright 2018 by Satoshi Okamoto

a schematic illustration of the implementation of MAMR [13]. An STO is placed between the write pole and the shield; the write field from the write pole and a microwave field from the STO are simultaneously applied to the medium. Writing a bit pattern then becomes possible even though the write field is not large enough. Prototypes of the MAMR head have been fabricated and their read/write performances have also been demonstrated [13–15]. Thus, MAMR is expected as a technology to increase the recording density of HDD further. Moreover, MAMR has the potential to achieve 3D recording, which can drastically increase recording density. To the extent that the recorded bits are vertically stacked, selective ***writing*** and ***reading*** are required in 3D magnetic recording to access the specific bit.

The objective of this chapter is to describe and explain recent research advances and prospects for MAMR and MAMR-based 3D magnetic recording.

10.2 Microwave-Assisted Magnetic Recording (MAMR)

10.2.1 Microwave-Assisted Switching (MAS)

Before delving into an explanation of microwave-assisted switching, it is instructive to examine first the basic physics of magnetization switching. Since magnetization is an angular momentum, the magnetization follows the equation of motion for angular momentum (Eq. 10.1),

$$\frac{\mathrm{d}\mathbf{m}}{\mathrm{d}t} = -\gamma[\mathbf{m} \times \mathbf{h}_{\mathrm{eff}}], \tag{10.1}$$

where $\mathbf{m}$ is the unit vector of magnetization, $\mathbf{h}_{\mathrm{eff}}$ is the effective magnetic field vector exerted on the magnetization, t is the time, and γ is the gyromagnetic ratio. Equation (10.1) shows that the change in angular momentum equals the torque force exerted on the angular momentum. Thus, magnetization precesses when it changes direction.

Since the precession frequency of the magnetic material used as a recording medium is in the microwave frequency range, the precessional motion of the magnetization in the medium is excited resonantly when a microwave field is applied. This precessional motion in the laboratory frame can be converted to a static motion in the rotating frame with the microwave frequency f_{rf}. This coordinate transformation from the laboratory frame to the rotating frame is given by Eq. (10.2):

$$\begin{aligned}\left[\frac{\mathrm{d}\mathbf{m}}{\mathrm{d}t}\right]' &= \frac{\mathrm{d}\mathbf{m}}{\mathrm{d}t} - [2\pi f_{rf}\mathbf{e}_z \times \mathbf{m}] \\ &= -\gamma\left[\mathbf{m} \times \left(\mathbf{h}_{\text{eff}} - h_{\omega}\mathbf{e}_z\right)\right],\end{aligned} \tag{10.2}$$

where $h_{\omega} = 2\pi f_{rf}/\gamma$. The last form of Eq. (10.2) has the same form as Eq. (10.1); however, the field term has an additional component of $-h_{\omega}\mathbf{e}_z$. This indicates that the precessional motion of the magnetization in the laboratory frame is equivalent to the fictitious field of $-h_{\omega}$ along the rotating axis in the rotating frame. Therefore, the field required to switch the magnetization becomes small owing to the presence of h_{ω}, and it decreases linearly with increasing f_{rf}. This is the reduction mechanism of the switching field in MAS [16].

The linear reduction behavior of the switching field H_{sw} in MAS is clearly confirmed by computer simulation (see Fig. 10.5) that was carried out by calculating the Landau–Lifshitz–Gilbert (LLG) equation for a single spin with an anisotropy field H_k of 20 kOe. The value of H_{sw} decreases linearly with increasing f_{rf} until a certain frequency. This frequency over which the microwave-assistance effect disappears is called the critical frequency [16], which reflects the notion that the magnetization precession cannot follow the very high microwave frequency. An analytical calculation shows that the magnetization precession becomes unstable at the critical frequency [9]. A more rigorous theoretical explanation of the MAS behavior may be found elsewhere [17, 18].

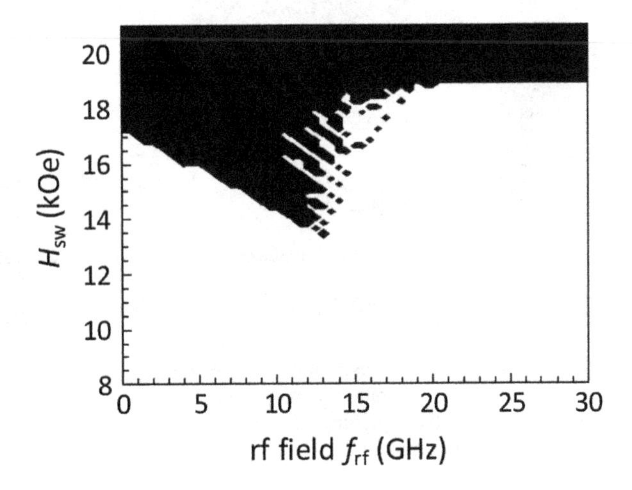

Fig. 10.5 Computer simulation result of MAS for a single spin model with anisotropy field H_k of 20 kOe. A circularly polarized microwave field of 0.5 kOe is applied. Copyright 2018 by Satoshi Okamoto

The experimental demonstration of MAS using a perpendicular magnetic dot was first reported in 2012 by Okamoto and coworkers [4]. Figure 10.6a shows a schematic illustration of the structure of the device used in this MAS experiment. The microwave field was generated by using a microscale conductor strip line, while the nanoscale magnetic dot was fabricated using the electron-beam lithography technique, and was placed underneath the strip line with an insulating layer inserted between them. The microwave was fed into the strip line from a signal generator, after which a transverse microwave field was applied to the magnetic dot.

Magnetization switching of the dot was detected electrically using the Anomalous Hall Effect (AHE) technique. Figure 10.6b shows the AHE curves of a single magnetic dot of 120 nm in diameter as a function of the direct current (dc) magnetic field H_{dc} under the application of microwaves with various f_{rf} [4]. The abrupt change in the AHE signal indicates the magnetization switching of the dot. It can clearly be seen that the switching field H_{sw} decreases significantly under the assistance of the microwaves. The f_{rf} dependence of H_{sw} is plotted in Fig. 10.6c, which shows that the H_{sw} obviously decreases linearly with increasing f_{rf} and abruptly increases at the critical frequency, as predicted by the simulation (see Fig. 10.5). Note that the H_{sw}

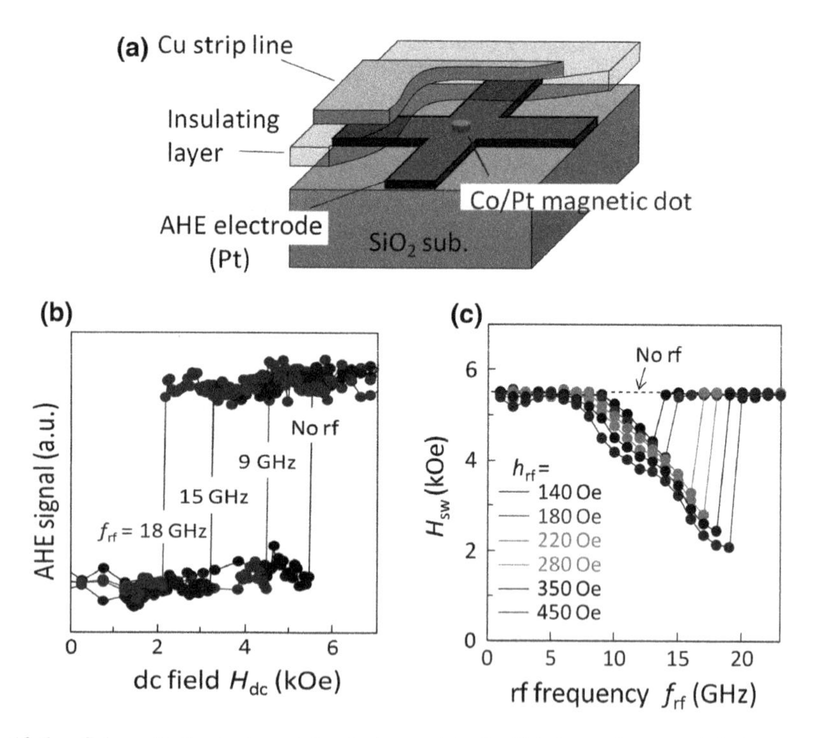

Fig. 10.6 **a** Schematic illustration of the device used in the MAS experiment. **b** AHE curves of a single Co/Pt dot of 120 nm in diameter under applying a microwave field of 450 Oe. **c** Frequency f_{rf} dependence of the switching field H_{sw} with various microwave field amplitude h_{rf}. Reproduced from Ref. [4]. Copyright 2012 by american physical society

reduction rate against f_{rf} is insensitive to the microwave field amplitude h_{rf}, whereas the critical frequency strongly depends on h_{rf}. In addition to these results, it is of great interest that the experimentally observed reduction of H_{sw} strongly depends on the dot diameter [5, 16].

Figures 10.7a and 10.7b show the MAS behavior of the dot array sample with diameter $D = 50$ nm and 230 nm, respectively. The reduction of H_{sw} for $D = 230$ nm is evidently much larger than that for $D = 50$ nm. Here, the minimum switching field $H_{\text{sw,min}}$ is defined as the value of H_{sw} at the critical frequency, and is plotted as a function of D in Fig. 10.7c. $H_{\text{sw,min}}$ decreases monotonically with increasing D, indicating that the MAS effect enhances for larger D. In the absence of microwave application ($H_{\text{sw}}^{\text{no rf}}$ in Fig. 10.7c), H_{sw} also decreases with increasing D, but its reduction is very limited. The critical frequency also exhibits the dot diameter dependence as shown

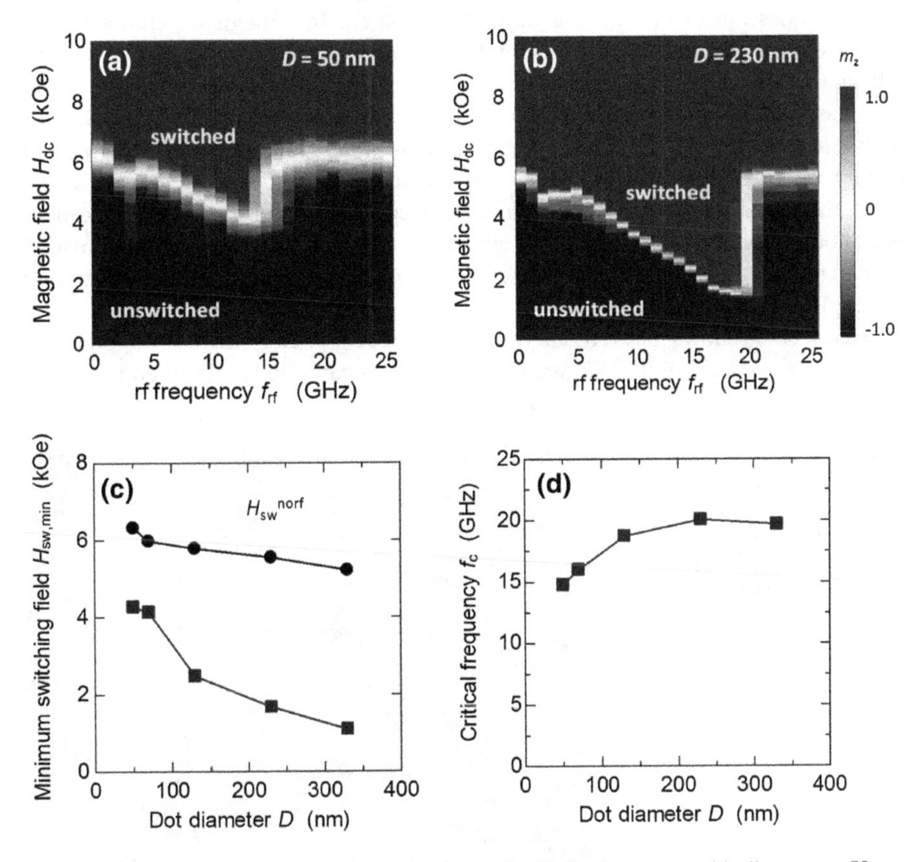

Fig. 10.7 Contour maps of the MAS behaviors of the Co/Pt dot arrays with diameter **a** 50 nm and **b** 230 nm, respectively, under the assistance of rf field $h_{\text{rf}} = 500$ Oe as functions of magnetic field H_{dc} (vertical axis) and rf frequency f_{rf} (abscissa axis). A color bar expresses the normalized AHE voltage. **c** Minimum switching field $H_{\text{sw,min}}$ and **d** critical frequency f_{c} as a function of dot diameter D. Circles denote the switching field in the absence of rf field $H_{\text{sw}}^{\text{no rf}}$. Reproduced from Ref. [5]. Copyright 2014 by AIP publishing

in Fig. 10.7d. According to detailed studies using micromagnetic simulation [5, 16], this enhancement of the MAS effect is explained by the higher-order spin-wave excitation in the dot, which is a non-uniform magnetization precession mode. Thus, it can be concluded that the MAS effect depends strongly on the magnetization precession mode.

Although the MAS behavior of dot samples can be described by the theories and computer simulations as explained above, the dot samples are quite different from the actual HDD recording medium. In fact, the HDD recording medium is a granular medium in which the magnetic nanoparticles are uniformly dispersed; an example is shown in Fig. 10.8 [6]. The MAS effect in a granular film has also been demonstrated experimentally [6, 19, 20]; however, its behavior is different from that observed in dot samples.

In most of the previous MAS experiments in magnetic granular films, the switching field does not exhibit a clear dependence on f_{rf}. In a magnetic granular film, there are finite interparticle magnetic interactions (exchange and dipole), as well as a large thermal agitation effect as a result of the nanoscale particle size. It is likely that these factors smear the MAS effect in a magnetic granular film [21]. Accordingly, in order to observe a clear MAS effect in a magnetic granular film, a large enough microwave field is required for overcoming the effects of the above factors. To realize this experimental condition, the device structure for the MAS experiment was thoroughly upgraded, following which the upgraded device structure provided a clearer MAS effect successfully observed in a magnetic granular film as illustrated in Fig. 10.9 [22].

Figure 10.9a shows the AHE curves of a magnetic granular film under application of a microwave field of 950 Oe with various f_{rf}. As can also be observed in the dot

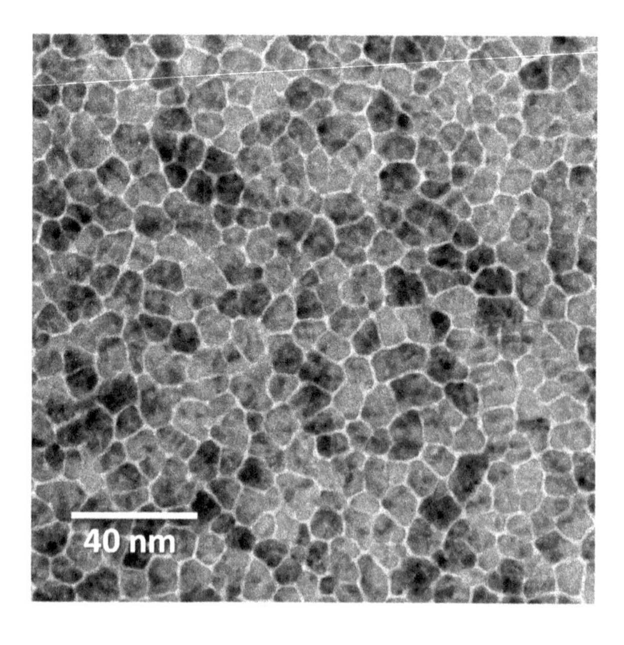

Fig. 10.8 Plan view of transmission electron microscopy image of $CoCrPtTiO_2$ film. Reproduced from Ref. [6]. Copyright 2013 by AIP publishing

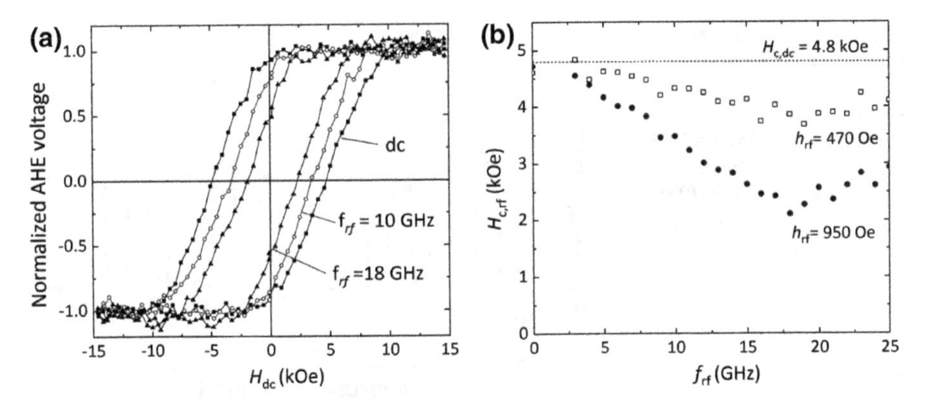

Fig. 10.9 **a** AHE curves of a CoCrPt-TiO_2 granular film under a microwave field of 950 Oe. **b** Frequency f_{rf} dependence of the coercive field H_c with various h_{rf}. Reproduced from Ref. [22]. Copyright 2018 by The Japan society of applied physics

sample shown in Fig. 10.6b, the AHE curve of the magnetic granular film clearly shifts toward the lower field region with increasing f_{rf}. Note that the AHE curve of the magnetic granular film exhibits a gradual change against the magnetic field, in contrast to the abrupt change seen in the dot sample. This is due to the gradual switching of the constituent nanoparticles in the magnetic granular film. Here, instead of the switching field H_{sw} that is well defined for the dot sample, the coercivity H_c is used for the magnetic granular film, which is defined as the magnetic field at the point of zero AHE signal on the AHE curve. It is plotted as a function of f_{rf}, as shown in Fig. 10.9b [22]. Under the assistance of a large amplitude microwave field, H_c of the granular film decreases linearly with f_{rf}. Although the H_{sw} reduction rate for the dot sample is insensitive to h_{rf} as seen in Fig. 10.6c, the H_c reduction rate for the magnetic granular film clearly depends on h_{rf}. According to Eq. (10.2), the reduction rate is inversely proportional to the gyromagnetic ratio γ, as verified experimentally and theoretically [16–18]. Moreover, the gradual increase of H_c in the frequency range over the critical frequency is also quite different from the abrupt increase in H_{sw} for the dot samples. These differences could be attributed to the interparticle magnetic interactions and/or to the thermal agitation effect on the magnetic granular film. To fully understand the MAS effect for the magnetic granular film, a more detailed study of these factors on the MAS effect is required.

The final topic to be introduced in this section is of great interest. The microwave field generated from a strip line is a linearly polarized wave, which is considered as a sum of clockwise (CW) and counterclockwise (CCW) circularly polarized waves. The chirality of the magnetization precession is uniquely determined by the magnetization direction of either up or down. Therefore, it is impossible to switch the magnetization direction only by applying a linearly polarized microwave field without applying a dc field, because the two components of circularly polarized waves (CW and CCW waves) promote equally magnetization switching from up to down and vice versa. Thus, the application of a dc field is required to break this equality,

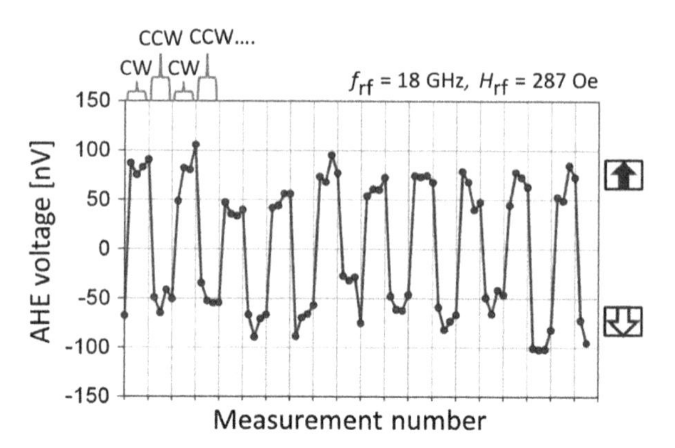

Fig. 10.10 Experimental demonstration of zero-dc-field MAS of a single Co/Pt dot by alternately applying CW and CCW microwave fields. Reproduced from Ref. [23]. Copyright 2017 by Springer Nature publishing AG

thereby realizing the deterministic MAS behavior. If a circularly polarized wave was applied, however, a deterministic MAS behavior would be expected without applying a dc field (that is, zero-dc-field MAS). This was well demonstrated using a device similar to that shown in Fig. 10.6a, except that two crossing strip lines were incorporated instead [23]. A careful tuning of the phase and amplitude of the microwaves following in these two strip lines allowed for the generation of a circularly polarized microwave at the crossing point. Consequently, zero-dc-field MAS behavior was successfully demonstrated as displayed in Fig. 10.10 [23]. The magnetization direction is deterministically switched by applying CW and CCW waves. This zero-dc-field MAS is not only expected to be of great importance to the future ultra-high density HDD applications but is also of great potential for solid-state memory applications.

10.2.2 Spin-Torque Oscillator (STO) for MAMR

A spin-torque oscillator (STO) is one of the spintronic devices, which can generate a microwave by passing a dc current through a nanoscale magnetic multilayer that consists of a so-called oscillation layer and a reference layer (RL) [24]. Since the current passing through the RL becomes spin polarized, the magnetization of the oscillation layer experiences a torque force because of the spin angular momentum transfer from the spin-polarized current. This is called as spin-transfer torque [25] and is also a key mechanism of the magnetic-random access memory (MRAM), which is expected to become the next-generation universal memory. In an STO, an auto-oscillation of magnetization precession in the oscillation layer is caused by the spin-transfer torque. STOs are expected to be applied in various fields and fulfill various roles. These include applications such as nanoscale microwave generators [26], highly sensitive magnetic sensors [27], and telecommunication devices [28], among others. When it is used for MAMR applications, a large amplitude microwave field can be generated by inducing a large cone angle magnetization precession in the oscillation layer. In this sense, the oscillation layer in STOs used in MAMR applications is usually referred to as a field generating layer (FGL).

Figure 10.11 shows an example of scanning electron microscope image of a prototype MAMR head with an STO, and Fig. 10.12 shows two typical layer stacking structures of STOs for MAMR application. They have perpendicularly magnetized RLs and soft magnetic FGLs with large saturation magnetization. One has a perpendicularly magnetized cap layer (PL) in contact with the FGL, which favors the magnetization of the FGL perpendicularly through the exchange coupling between them [2, 29, 30]. The layer structure of this type of STO is schematically shown in Fig. 10.12a. According to some computer simulations based on the LLG equation, this type of STO can generate h_{rf} of a few kOe with f_{rf} of a few tens GHz, which satisfies the requirements for MAMR applications. However, this type of STO consists of at least four layers including an interlayer between the RL and the FGL, which makes the gap between the shields and the write pole wider. This wide gap may degrade the recording performance of the HDD owing to the broad write field distribution. On the other hand, other types of STOs without using the PL are also proposed for MAMR applications [30–32] as schematically shown in Fig. 10.12b. This type of STO can make the gap between the shields and the write pole narrower in comparison with the former. According to the computer simulation, this type of STO can also generate h_{rf} of a few kOe with f_{rf} of the order of tens GHz. It also achieves a wider range of a stable oscillation condition. The large cone angle magnetization precession of this type of STO has also been confirmed experimentally [12].

The MAS was successfully demonstrated experimentally in 2017 by Suto and coworkers [33] using the microwave field generated by an STO. This experiment showed that using an STO the reduction rate of H_{sw} is almost identical to that observed

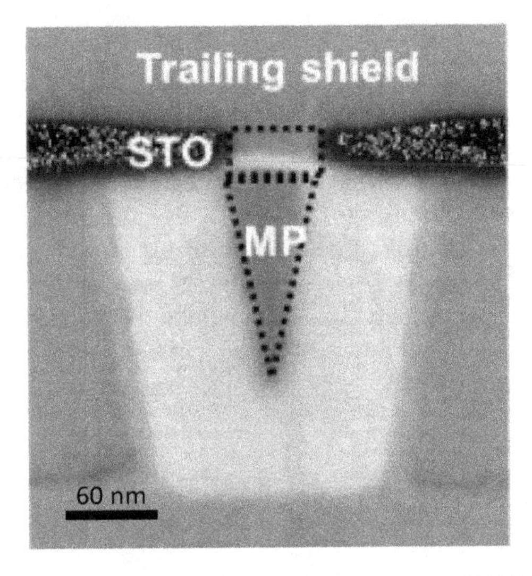

Fig. 10.11 Scanning electron microscopy image of the prototype MAMR head. MP is the main write pole, and STO is placed between MP and trailing shield. Copyright 2018 by Ikuya Tagawa

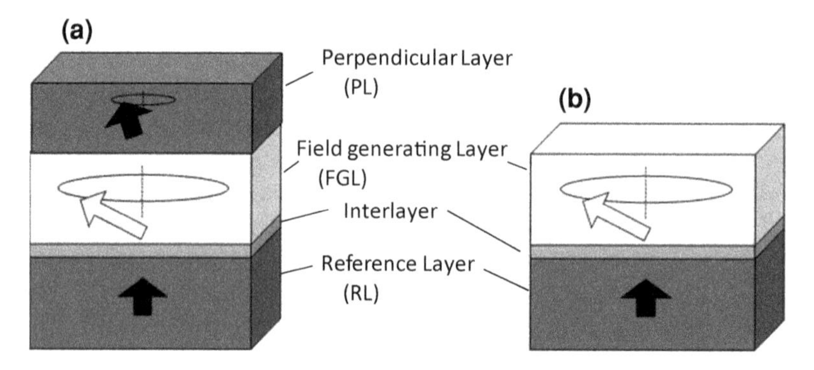

Fig. 10.12 Schematic illustrations of two types of STO for MAMR application. Copyright 2018 by Satoshi Okamoto

when using the microwave field generated by a strip line. Thus, that the standalone STO works well as a microwave field generator for MAMR was verified. On the other hand, the STO in the prototype MAMR head (Fig. 10.11) gave a read/write performance which was not as good as expected from the computer simulation and standalone STO performance [15]. Since the STO in the MAMR head is surrounded by the magnetic materials, such as the write pole and shield, further optimization is required for the stable microwave generation from the STO in the MAMR head.

10.3 MAMR-Based 3D Magnetic Recording

As mentioned in the preceding section, H_{sw} in MAS decreases linearly with increasing f_{rf} and abruptly increases at the critical frequency. This behavior leads to the expectation that the layer-selective MAS using multiple stacked media possess different f_{rf} dependencies [34–39]. This concept was proposed soon after the first report of MAMR. However, it is difficult to read the signal from each layer of the multiply-stacked media by using a conventional field sensor type read head. To solve this problem, an STO can be used as a frequency-resolvable read head [40, 41]. Thus, the layer-selective MAS and frequency-resolvable STO reader allow for the realization of 3D magnetic recording, which has the potential to enhance drastically the recording density of HDDs. In the following section, present developments and technical issues for the layer-selective MAS and STO reader are described and discussed.

10.3.1 Layer-Selective MAS

Figure 10.13 shows a concept image of the layer-selective MAS [16]. By tuning the value of f_{rf} and the dc field, a layer-selective MAS is possible using multiply-stacked media. Many computer simulations of the layer-selective MAS have been reported so far [33–36]. The experimental demonstration of the layer-selective MAS, as shown in Fig. 10.14 [38], was first reported in 2016. In this experiment, a dot with double-magnetic layers was used, and the switching of each layer was detected using the magnetoresistance effect. The lower-layer (LL) is magnetically harder than the upper-layer (UL). Therefore, the switching order of these layers in the absence of a microwave field was the UL switching first and the LL switching second with increasing the dc field. When the microwave field was applied, the switching fields of both the LL and UL decreased with increasing the microwave frequency f_{rf}; however, their switching order did not change until $f_{rf} < 15$ GHz. Both layers switched simultaneously at $f_{rf} = 15$ GHz, and then the switching order changed

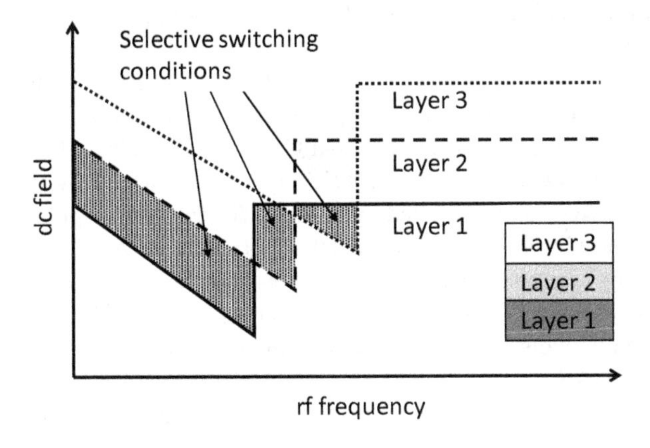

Fig. 10.13 Illustration of layer-selective MAS concept. Copyright 2018 by Satoshi Okamoto

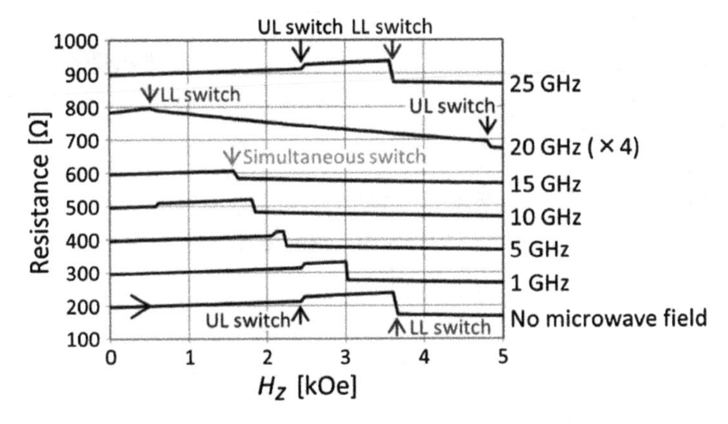

Fig. 10.14 Experimental demonstration of layer-selective MAS for a dot of double-magnetic layers. The lower-layer (LL) is magnetically harder than the upper-layer (UL). Reproduced from Ref. [38]. Copyright 2016 by American physical society

at $f_{\mathrm{rf}} = 20$ GHz. At this point, the UL switched first and the LL second. These experimental results confirm the layer-selective MAS for the double-magnetic layer dot.

Here, it should be mentioned that the Eigen frequency of a ferromagnet, i.e., the ferromagnetic resonance frequency f_{FMR}, is proportional to the effective field exerted on it. This means that f_{FMR} of each layer varies depending on the magnetization direction of the adjacent layer due to dipolar interaction. This shift in f_{FMR} of each layer is a serious problem for the operation of STO reading. To prevent this problem, an antiferromagnetically coupled (AFC) medium has been proposed for 3D magnetic recording [42–44]. An AFC layer consists of two magnetic layers with a non-magnetic specific interlayer; the magnetic layers favor their magnetization direction antiparallel due to the antiferromagnetic coupling through the interlayer. Therefore, since an AFC layer basically has no stray field, there is no dipolar interaction in multiply-stacked AFC layers. Thus, the frequency shift problem mentioned above can be avoided by using AFC layers.

The layer-selective MAS for a dot of double AFC layers was also successfully demonstrated, as shown in Fig. 10.15 [35]. The sample was composed of magnetically soft and hard AFC layers, referred to as s-AFC and h-AFC in Fig. 10.15, respectively. The magnetization switching of each AFC layer was detected by the AHE effect; their switchings were discriminated from the different AHE signal amplitude. As well as for the case in Fig. 10.14, the switching order was s-AFC switching first and h-AFC switching second for $f_{\mathrm{rf}} \leq 18$ GHz. This switching order changed to h-AFC switching first and s-AFC switching second at $f_{\mathrm{rf}} = 18.9$ GHz. As f_{rf} increased further, the simultaneous switching of both AFC layers occurred at $f_{\mathrm{rf}} = 19.5$ GHz, after which the MAS effect of both AFC layers disappeared for $f_{\mathrm{rf}} \geq 20$ GHz. Thus, the layer-selective MAS was also confirmed experimentally in a dot of double AFC layers.

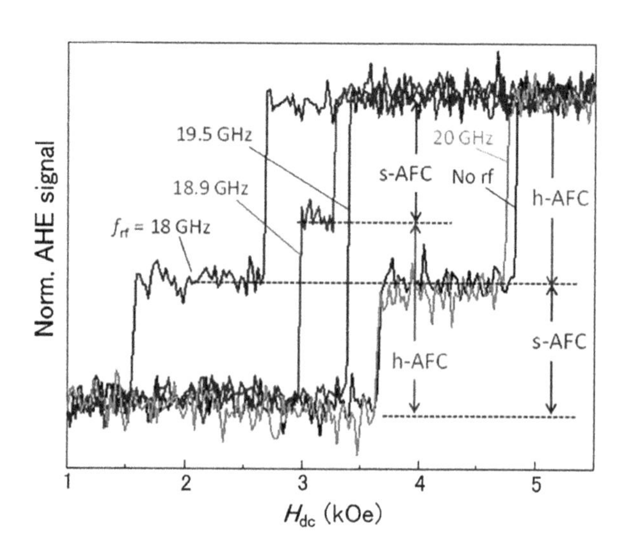

Fig. 10.15 Experimental demonstration of layer-selective MAS for a dot of double AFC layers. s-AFC and h-AFC denote the soft and hard AFC layers, respectively. Reproduced from Ref. [39]. Copyright 2018 by AIP publishing

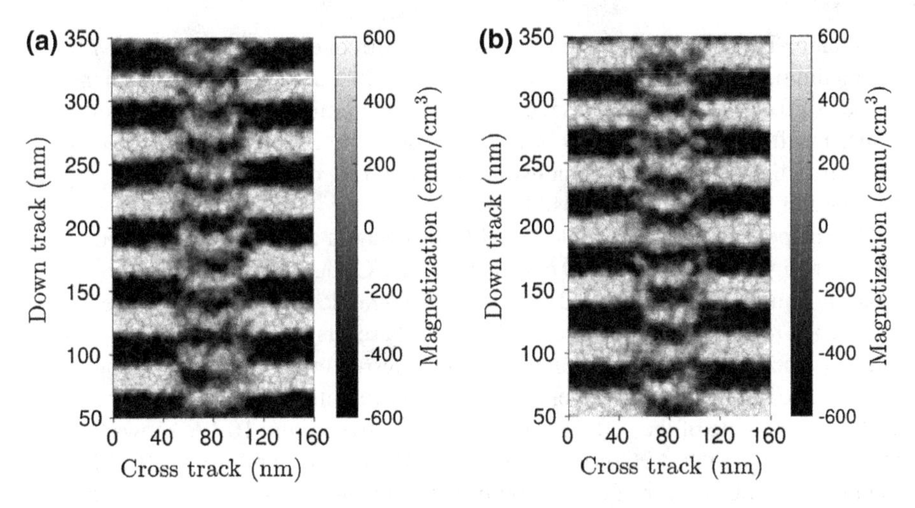

Fig. 10.16 Images of written tracks for **a** single layer (SL) granular media and **b** AFC granular media. Images are averages of ten tracks. 15 nm bit length and 1693 kfci. Copyright 2018 by Simon Greaves

It should be noted that the layer-selective switching was realized only in a very narrow frequency range for both the simple double layer stacked dot sample (Fig. 10.14) and the double AFC layer stacked dot sample (Fig. 10.15). Moreover, the simultaneous switching also occurred in both samples. At present, the reasons for these behaviors are not clear. Some unknown factors associated with the interlayer interaction may be at work.

Thus, the AFC layer is an important technology for the 3D magnetic recording, but the AFC layer is unexpectedly also very effective in the conventional MAMR. Figure 10.16 shows the computer simulation results for written track pattern images of the usual single layer (SL) granular medium and the AFC type granular medium [44]. The bit transition and the bit shape of the AFC granular medium can be seen much clearer than those of the SL granular medium. This is because interparticle dipolar interaction in the AFC granular medium is very small. This very clear written track pattern in the AFC granular medium demonstrates the low noise recording performance.

10.3.2 STO Reader for 3D Recording

An STO reader has been proposed for 3D magnetic recording [40, 41], which can detect the recorded signal of the AFC medium in the frequency domain. Since the AFC medium has very low or zero stray field, a conventional field sensor type read head is not applicable. One of the important features of an STO is that the oscillation frequency is tunable by changing the current density and/or the bias magnetic field amplitude. Therefore, when the STO oscillation frequency is in tune with f_{FMR} of the medium, the STO and the medium resonate with each other. This resonant state causes shifts of frequency, phase, and oscillation amplitude in the STO oscillation. Thus, a shift in these parameters is available for the detection of the recorded signal. Moreover, the experimental demonstration of the STO reader reveals that a very fast response of less than 1 ns is possible [45].

The response of an STO reader to AFC dots was calculated [40] using the model shown in Fig. 10.17a for the STO reader and the AFC dot. The AFC dot was composed of antiferromagnetically coupled soft and hard magnetic layers. The hard layer is used to store the recorded data, and the soft layer easily reacts with the STO. The soft layer takes up and down magnetization directions, referred to as (u) and (d), respectively, as indicated in Fig. 10.17b. Under the calculation conditions used, the STO resonates with only the state of the soft layer magnetization (d). Figure 10.17c shows the STO response against the AFC dot row of (d)(u)(d)(u)(d)(u), while Fig. 10.17d shows the instantaneous power of oscillation of the STO and the AFC dots. It can be seen

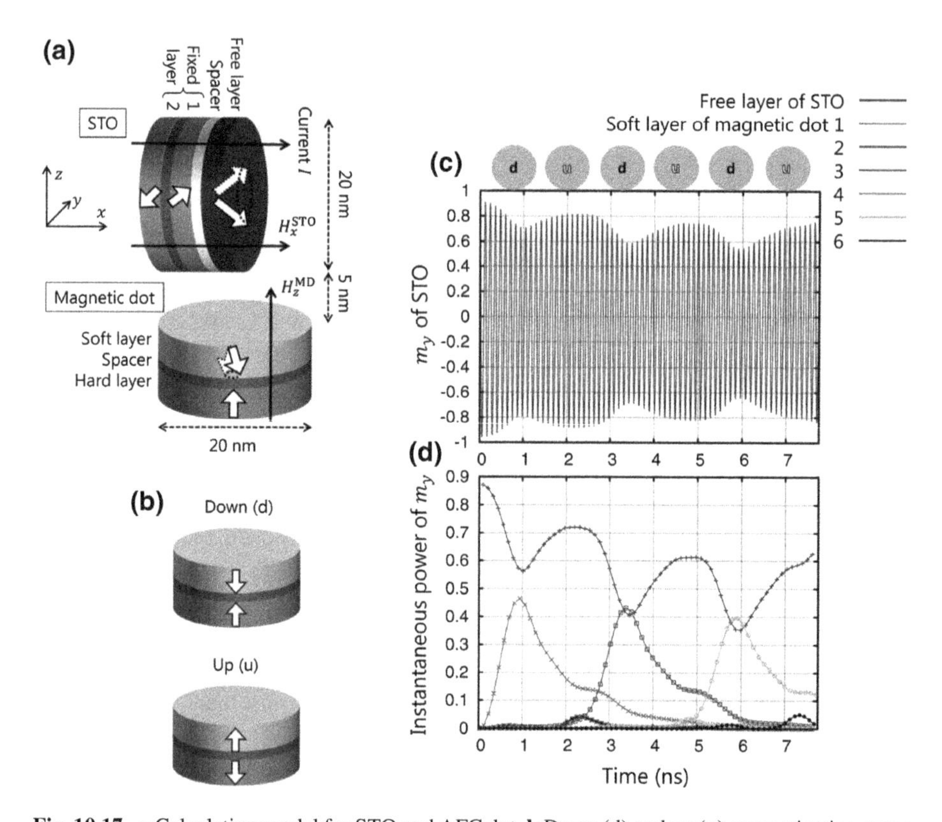

Fig. 10.17 **a** Calculation model for STO and AFC dot. **b** Down (d) and up (u) magnetization states of AFC dots, **c** STO oscillation response to the AFC dot row of (d)(u)(d)(u)(d)(u), **d** Instantaneous power of oscillation of STO and AFC dots. Reproduced from Ref. [40]. Copyright 2018 by AIP publishing

clearly that the STO oscillation amplitude is damped when it moves over the AFC dot of the state (d). At the same time, the soft layer magnetization of the AFC dot of the state (d) starts to oscillate largely. On the other hand, the STO oscillation recovers once it moves over the AFC dot of the state (u), and the reaction of the AFC dot of the state (u) is very small. This demonstrates that the STO reader can read out a recorded signal from the AFC medium, even though the stray field from the AFC medium is negligibly small.

Since the signal waveform from the STO reader is quite different from that of a conventional field sensor type HDD reader, a decoding method is required to reproduce the recorded data from the STO reader signal waveform. To develop a decoding method for the STO reader signal, a large amount of signal wave data is required. Unfortunately, the available signal waveform, which is obtained by computer simulation (see Fig. 10.17) is too short. To solve this problem, a mathematical model to produce a signal waveform from the STO reader has been developed first.

Figure 10.18a shows an example of the signal waveform from the STO reader in response to the recorded data pattern of "000111000111" [46]. Here, "0" and "1" in Fig. 10.18 correspond to (d) and (u) in Fig. 10.17, respectively. As mentioned above, the signal wave from the STO reader is a sinusoidal GHz frequency wave with the recorded data pattern imprinted in the form of amplitude modulation. Therefore, the envelope of the signal wave is available for the decoding process. This envelope pattern can easily be reproduced mathematically by convoluting the attenuation function, as shown by the solid line in Fig. 10.18a. Thus, the mathematically produced envelope pattern can be analyzed, and a bit error rate (BER) for the STO reader can be evaluated. The signal includes the computationally generated white noise and the recorded bit position fluctuation σ_x. Both differential detection for the envelope pattern and a soft-output Viterbi algorithm (SOVA) [47] as a maximum likelihood decoder for a partial response (PR) [48] channel are used to study the signal. The evaluated BER performances are plotted in Fig. 10.18b. It can clearly be seen that the SOVA detection exhibits much better BER performance than that of the differential detection. The good BER performance of the SOVA detection indicates that the STO shows good potential as a reader device for 3D magnetic recording using AFC media.

10.4 Summary

Recent advances and prospects for microwave-assisted magnetic recording (MAMR) and MAMR-based 3D magnetic recording have been presented and described. At present, MAMR has been highly expected as the next-generation ultra-high-density magnetic recording technology. It should be stressed that MAMR is not only a simple technology of extending the present recording density limits, but it will also lead to a qualitative technological change. In conventional magnetic recording, the writing and reading operations are performed by means of the field strength. Contrary to this conventional operation, the writing in MAMR and the sensing in the STO reader are performed in the frequency domain. These frequency-domain writing and reading

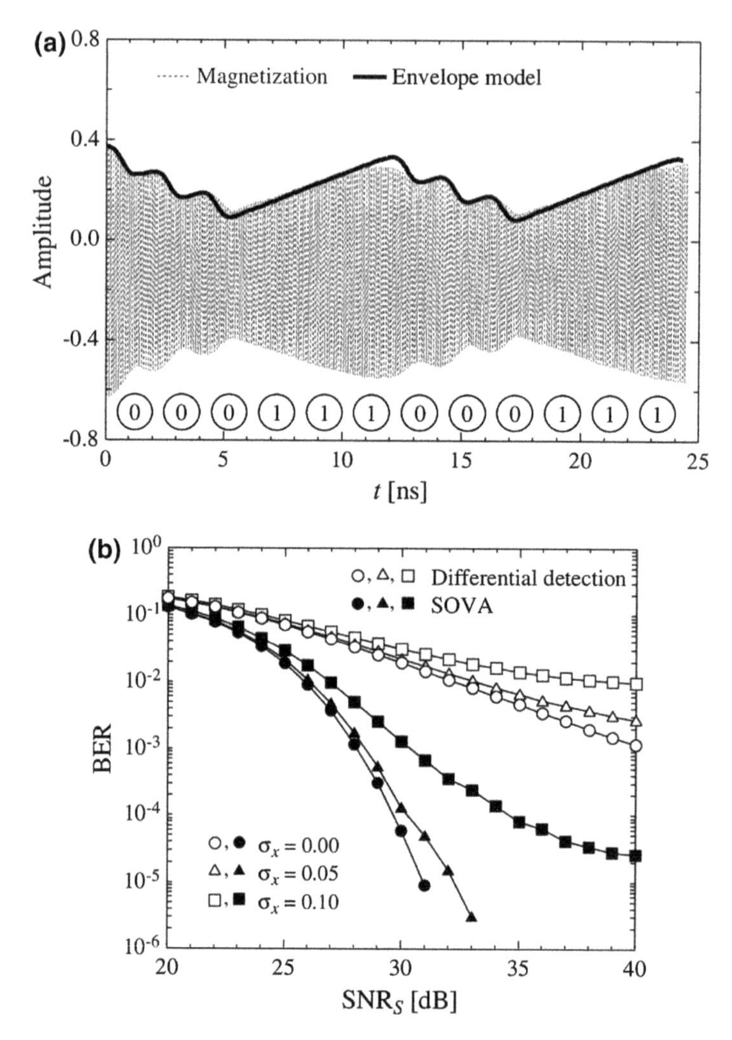

Fig. 10.18 **a** Envelope model for the signal waveform of the STO reader against the recorded data pattern of "000111000111". **b** Bit error rate (BER) performance of differential detection and SOVA for the signal of STO with white noise and recorded bit position fluctuation σ_x. Copyright 2018 by Yasuaki Nakamura

technologies will enable the development of 3D magnetic recording, which is difficult to realize in the framework of conventional magnetic recording systems. Moreover, many other relevant technologies have emerged from the studies on MAMR and MAMR-based 3D magnetic recording, such as zero-dc-field MAS, low noise recording operation in AFC media, and very high-speed operation of STO sensing, among others. These are expected to be applied not only in magnetic recording but also in other fields of applications, such as solid-state memory and sensor technology. In addition to these technological advances, MAMR and MAMR-based 3D magnetic recording strongly stimulate frontier studies on high-frequency and nonlinear spin dynamics. Present understandings within these fields are not yet broad enough and many unexplored subjects remain. Advances in these research fields will strongly accelerate the development of MAMR and 3D magnetic recording and will also promote the development of other novel technologies and devices.

References

1. Shiroishi Y, Fukuda K, Tagawa I, Iwasaki H, Takenoiri S, Tanaka H, Mutoh H, Yoshikawa N (2009) Future options for HDD storage. IEEE Trans Magn 45:3816–3822
2. Zhu J-G, Zhu X, Tang Y (2008) Microwave assisted magnetic recording. IEEE Trans Magn 44:125–131
3. Thirion C, Wernsdorfr W, Mailly D (2003) Switching of magnetization by nonlinear resonance studied in single nanoparticles. Nat Mater 2:524–527
4. Okamoto S, Kikuchi N, Furuta M, Kitakami O, Shimatsu T (2012) Switching behaviors and its dynamics of a Co/Pt nanodot under the assistance of rf fields. Phys Rev Lett 109:237209
5. Furuta M, Okamoto S, Kikuchi N, Kitakami O, Shimatsu T (2014) Size dependence of magnetization switching and its dispersion of Co/Pt nanodots under the assistance of radio frequency fields. J Appl Phys 115:133914
6. Okamoto S, Kikuchi N, Hotta A, Furuta M, Kitakami O, Shimatsu T (2013) Microwave assistance effect on magnetization switching in Co-Cr-Pt granular film. Appl Phys Lett 103:202405
7. Nozaki Y, Matsuyama K (2006) Numerical study for ballistic switching of magnetization in single domain particle triggered by a ferromagnetic resonance within a relaxation time limit. J Appl Phys 100:053911
8. Okamoto S, Kikuchi N, Kitakami O (2008) Magnetization switching behavior with microwave assistance. Appl Phys Lett 93:102506
9. Okamoto S, Igarashi M, Kikuchi N, Kitakami O (2010) Microwave-assisted switching mechanism and its stable switching limit. J Appl Phys 107:123914
10. Tang Y, Zhu J-G (2008) Narrow track confinement by AC field generation layer in microwave-assisted magnetic recording. IEEE Trans Magn 44:3376–3379
11. Masuko J, Matsubara M, Hashimoto J, Kanai H, Uehara Y, Ibusuki T, Sato M, Wada T, Suzuki Y (2009) Microwave oscillations of the giant magnetoresistive element in a magnetic field perpendicular to the plane. IEEE Trans Magn 45:3430–3433
12. Bosu S, Sepehri-Amin H, Sakuraba Y, Kasai S, Hayashi M, Hono K (2017) High frequency out-of-plane oscillation with large cone angle in mag-flip spin torque oscillators for microwave assisted magnetic recording. Appl Phys Lett 110:142403
13. Shiimoto M (2010) Experimental feasibility of spin-torque oscillator with synthetic field generation layer for microwave assisted magnetic recording. In: 55th MMM conference, DF-10 Atlanta, 14–18 Nov 2010
14. Takeo A (2014) MAMR R/W performance improvement by mag-flip STO assist. In: Intermag. Conference, AD-2 Dresden, 4–8 May 2014
15. Tagawa I, Shiimoto M, Matsubara M, Nosaki S, Urakami Y, Aoyama J (2016) Advantage of MAMR read-write performance. IEEE Trans Magn 52:3101104
16. Okamoto S, Kikuchi N, Furuta M, Kitakami O, Shimatsu T (2015) Microwave assisted magnetic recording technologies and related physics. J Phys D Appl Phys 48:353001
17. Taniguchi T (2014) Magnetization reversal condition for a nanomagnet within a rotating magnetic field. Phys Rev B 90:024424
18. Suto H, Kudo K, Nagasawa T, Kanao T, Mizushima K, Sato R, Okamoto S, Kikuchi N, Kitakami O (2015) Theoretical study of thermally activated magnetization switching under microwave assistance: switching paths and barrier height. Phys Rev B 91:094401

19. Nozaki Y, Ishida N, Soeno Y, Sekiguchi K (2012) Room temperature microwave- assisted recording on 500-Gbpsi-class perpendicular medium. J Appl Phys 112:083912
20. Boone CT, Katine JA, Marinero EE, Pisana S, Terris BD (2012) Demonstration of microwave assisted magnetic reversal in perpendicular media. J Appl Phys 111:07B907
21. Okamoto S, Kikuchi1 N, Kitakami1 O, Shimatsu T (2017) Influence of intergrain interactions and thermal agitation on microwave-assisted magnetization switching behavior of granular magnetic film. Appl Phys Express 10:023004
22. Kikuchi N, Shimada K, Shimatsu T, Okamoto S, Kitakami O (2018) Frequency dependence of microwave-assisted switching in CoCrPt granular perpendicular media. Jpn J Appl Phys 57:09TE02
23. Suto H, kanao T, Nagasawa T, Mizushima K, Sato R (2017) Zero-dc-field rotation-direction dependent magnetization switching induced by a circularly polarized microwave magnetic field. Sci Rep 7:13804
24. Kiselev SI, Sankey JC, Krivorotov IN, Emley NC, Schoelkopf RJ, Buhrman RA, Ralph DC (2003) Microwave oscillations of a nanomagnet driven by a spin-polarized current. Nature 425:380–383
25. Slonczewski JC (1996) Current-driven excitation of magnetic multilayers. J Magn Magn Mater 159:L1–L7
26. Kaka S, Pufall MR, Rippard WH, Silva TJ, Russek SE, Katine JA (2005) Mutual phase-locking of microwave spin torque nano-oscillators. Nature 437:389–392
27. Kudo K, Nagasawa T, Mizushima K, Suto H, Sato R (2010) Numerical simulation on temporal response of spin-torque oscillator to magnetic pulses. Appl Phys Express 3:043002
28. Choi HS, Kang SY, Cho SJ, Oh I-Y, Shin M, Park H, Jang C, Min B-C, Kim S-I, Park S-Y, Park CS (2014) Spin nano-oscillator-based wireless communication. Sci Rep 4:5486
29. Zhu J-G, Wang Y (2010) Microwave assisted magnetic recording utilizing perpendicular spin torque oscillator with switchable perpendicular electrodes. IEEE Trans Magn 46:751–757
30. Sato Y, Sugiura K, Igarashi M, Watanabe K, Shiroishi Y (2013) Thin spin-torque oscillator with high AC-field for high density microwave-assisted magnetic recording. IEEE Trans Magn 49:3632–3635
31. Igarashi M, Suzuki Y, Sato Y (2010) Oscillation feature of planar spin-torque oscillator for microwave-assisted magnetic recording. IEEE Trans Magn 46:3738–3841
32. Yoshida K, Yokoe M, Ishikawa Y, Kanai Y (2010) Spin torque oscillator with negative magnetic anisotropy materials for MAMR. IEEE Trans Magn 46:2466–2469
33. Suto H, Kanao T, Nagasawa T, Kudo K, Mizushima K, Sato R (2017) Switching field reduction of a perpendicular magnetic nanodot in a microwave magnetic field emitted from a spin-torque oscillator. Appl Phys Lett 110:132403
34. Winkler G, Suess D, Lee JJ, Fidler Bashir MA, Dean J, Goncharov A, Hrkac G, Bance S, Schrefl T (2009) Microwave-assisted three-dimensional multilayer magnetic recording. Appl Phys Lett 94:232501
35. Li S, Livshitz B, Bertram HN, Fullerton EE, Lomakin V (2009) Microwave-assisted magnetization reversal and multilevel recording in composite media. J Appl Phys 105:07B909
36. Tanaka T, Otsuka Y, Furomoto Y, Matsuyama K, Nozaki Y (2013) Selective magnetization switching with microwave assistance for three-dimensional magnetic recording. J Appl Phys 113:143908
37. Greaves S, Kanai Y, Muraoka H (2017) Multiple layer microwave-assisted magnetic recording. IEEE Trans Magn 53:3000510
38. Suto H, Nagasawa T, Kudo K, Kanao T, Mizushima K, Sato R (2016) Layer-selective switching of a double-layer perpendicular magnetic nanodot using microwave assistance. Phys Rev Appl 5:014003
39. Lu Y, Okamoto S, Kikuchi N, Kitakami O, Shimatsu T (2018) Layer-selective microwave-assisted magnetization switching in a dot of double antiferromagnetically coupled (AFC) layers. Appl Phys Lett 112:162404
40. Kanao T, Suto H, Kudo K, Nagasawa T, Mizushima K, Sato R (2018) Transient magnetization dynamics of spin-torque oscillator and magnetic dot coupled by magnetic dipolar interaction: reading of magnetization direction using magnetic resonance. J Appl Phys 123:043903

41. Suto H, Nagasawa T, Kudo K, Mizushima K, Sato R (2014) Nanoscale layer-selective readout of magnetization direction from a magnetic multilayer using a spin-torque oscillator. Nanotechnology 25:245501
42. Yang T, Suto H, Nagasawa T, Kudo K, Mizushima K, Sato R (2013) Readout method from antiferromagnetically coupled perpendicular magnetic recording media using ferromagnetic resonance. J Appl Phys 114:213901
43. Nakayama Y, Kusanagi Y, Shimatsu T, Kikuchi N, Okamoto S, Kitakami O (2016) Microwave-assistance effect on magnetization switching in antiferromagnetically coupled CoCrPt granular media. IEEE Trans Magn 52:3201203
44. Greaves S, Kanai Y, Muraoka H (2018) Antiferromagnetically coupled media for microwave-assisted magnetic recording. IEEE Trans Magn 54:3000111
45. Suto H, Nagasawa T, Kudo K, Mizushima K, Sato R (2011) Real-time measurement of temporal response of a spin-torque oscillator to magnetic pulses. Appl Phys Express 4:013003
46. Nakamura Y, Nishikawa M, Osawa H, Okamoto Y, Kanao T, Sato R (2018) Envelope detection using temporal magnetization dynamics of resonantly interacting spin-torque oscillator. AIP Adv 8:056512
47. Hagenauer J, Hoeher P (1989) A viterbi algorithm with soft-decision output and its applications. In: Proceedings of IEEE GLOBECOM, Dallas, TX, USA, 27–30 Nov 1989, pp 1680–1686
48. Kretzmer E.R (1966) Generalization of a technique for binary data communications. IEEE Trans Commun Technol COM-14, 67–68

www.ingramcontent.com/pod-product-compliance
Lightning Source LLC
Chambersburg PA
CBHW052214300125
21175CB00003B/41